Lecture Notes in Mathematics

1575

Editors:
A. Dold, Heidelberg
B. Eckmann, Zürich
F. Takens, Groningen

Marius Mitrea

Clifford Wavelets, Singular Integrals, and Hardy Spaces

Springer-Verlag

Berlin Heidelberg New York
London Paris Tokyo
Hong Kong Barcelona
Budapest

Author

Marius Mitrea
Institute of Mathematics
of the Romanian Academy
P. O. Box 1-764
RO-70700 Bucharest, Romania
and
Department of Mathematics
University of South Carolina
Columbia, SC 29208, USA

Mathematics Subject Classification (1991): 30G35, 42B20, 42B30, 31B25

ISBN 3-540-57884-6 Springer-Verlag Berlin Heidelberg New York
ISBN 0-387-57884-6 Springer-Verlag New York Berlin Heidelberg

CIP-Data applied for

© Springer-Verlag Berlin Heidelberg 1994
Printed in Germany

SPIN: 10130077   46/3140-543210 - Printed on acid-free paper *to Dorina*

Table of Contents

Page

Introduction IX

Chapter 1: Clifford Algebras 1

§1.1 Real and complex Clifford algebras 1

§1.2 Elements of Clifford Analysis 5

§1.3 Clifford modules 11

Chapter 2: Constructions of Clifford Wavelets 16

§2.1 Accretive forms and accretive operators 17

§2.2 Clifford Multiresolution Analysis. The abstract setting 18

§2.3 Bases in the wavelet spaces 23

§2.4 Clifford Multiresolution Analyses of $L^2(\mathbb{R}^m) \otimes \mathbb{C}_{(n)}$ 26

§2.5 Haar Clifford wavelets 30

Chapter 3: The L^2 Boundedness of Clifford Algebra Valued Singular Integral

Operators 42

§3.1 The higher dimensional Cauchy integral 43

§3.2 The Clifford algebra version of the $T(b)$ theorem 53

Chapter 4: Hardy Spaces of Monogenic Functions 60

§4.1 Maximal function characterizations 61

§4.2 Boundary behavior 70

§4.3 Square function characterizations 73

§4.4 The regularity of the Cauchy operator 82

Chapter 5: Applications to the Theory of Harmonic Functions 87

§5.1 Potentials of single and double layers 87

§5.2 L^2-estimates at the boundary 90

§5.3 Boundary value problems for the Laplace operator in Lipschitz domains 93

§5.4 A Burkholder-Gundy-Silverstein type theorem for monogenic functions
and applications 98

References 106

Notational Index 113

Subject Index 114

Introduction

As the seminal work of Zygmund [Zy] describes the state of the art in the mid 30's, much of classical Fourier Analysis, dealing with the boundary behavior of harmonic functions in the unit disc or the upper-half plane, has initially been developed with the aid of complex-variable methods. The success of extending these results to higher dimensions, the crowning achievement of Zygmund, Calderón and their collaborators, was largely conditioned upon devising new techniques, this time of purely real-variable nature (see e.g. [St], [To]).

Then why Clifford algebras? The basic motivation is that within this algebraic framework we can still do some sort of "complex analysis" in $\mathbb{R}^n$, for any n, which turns out to be much better suited for studying harmonic functions, say, than Several Complex Variables. For instance, any harmonic function is the real part of a Clifford analytic one and, on the operator side, the double layer Newtonian potential operator is the real part of the Clifford-Cauchy integral. At the heart of the matter lies the fact that while in general the square root of the Laplacian $\Delta = \partial_1{}^2 + \cdots + \partial_n{}^2$ is only a pseudo-differential operator in $\mathbb{R}^n$ with its usual structure, by first embedding $\mathbb{R}^n$ into a Clifford algebra we can do better than this, and realize $\Delta^{1/2}$ as a first order, elliptic differential operator, of Cauchy-Riemann type (though, a Clifford algebra coefficient one). The hypercomplex function theory has a long history, and its modern fundamentals have been laied down by Moisil, Teodorescu and Fueter (among others). However, much of the current research going on along these lines originates in the work of Coifman, McIntosh and their collaborators.

This book is conceived as a brief, fairly elementary and reasonably self-contained account of some recent developments in the direction of using Clifford algebra machinery in connection with relevant problems arising at the interface between Harmonic Analysis and Partial Differential Equations. Our goal is to provide the reader with a body of techniques and results which are of a general interest for these areas. Strictly speaking, there are no essentially new results, although perhaps some proofs appear for the first time in the literature. Yet, we believe that this presentation is justified by the point of view we adopt here. The text is by no means intended to be exhaustive and the topics covered rather reflect the interests and the limitations of the author. No specific knowledge of the subject is expected of the reader, although some familiarity with basic elements of classical Harmonic Analysis will help.

The plan of the book is as follows. Chapter 1 contains some preparatory material about Clifford algebras and Clifford analysis. The presentation is as concise as possible, yet aimed to give a sufficiently rich background for understanding the algebraic formalism used throughout. More detailed accounts on these matters can be found in [BDS] and [GM2].

The scope of the next two chapters is to treat Clifford algebra valued singular integral operators. The underlying idea for proving L^2−boundedness results ([Tc], [CJS]; cf. also [Dav2]) is fundamentally very simple. It consists of representing the given operator as an infinite matrix with respect to a certain Clifford algebra valued Riesz bases in L^2, whose specific properties ensure that this matrix has an almost diagonal form, i.e. the entries die fast enough off the main diagonal. Then a familiar argument based on Schur's lemma yields the result.

In Chapter 2 we construct such Riesz bases with wavelet structure, called Clifford wavelets, adapted to some Clifford algebra valued measures in $\mathbb{R}^n$, and having a priori prescribed properties of smoothness, cancellation and decay. In particular, to deal with the higher dimensional Cauchy singular integral operator on a Lipschitz hypersurface Σ in $\mathbb{R}^n$, we produce a (Clifford-)weighted Haar system which incorporates the information concerning the geometry of Σ (cf. [CJS], [AJM]). Once this is accomplished, one can work directly on Σ just as easily as if it were flat. Consequently, the L^2−boundedness of the higher dimensional Cauchy integral operator on Σ follows exactly as in the more classical case of the Hilbert transform in $\mathbb{R}$ (see e.g. [Ch]). This is worked out in detail in Chapter 3. Here we also outline the proof of the L^2−boundedness for a more general class of Clifford algebra valued singular integral operators satisfying the hypotheses of a Clifford T(b) theorem. This is done in the same spirit as before, i.e. essentially as a corollary of the existence of some suitable bases of Clifford wavelets.

A natural setting for studying the boundedness, regularity and boundary behavior of the Clifford-Cauchy integral on Lipschitz domains is a type of Hardy-like spaces of Clifford analytic functions which we discuss in Chapter 4. There is an interesting connection between these and the classical H^p spaces as introduced by Stein and Weiss [SW1] in that any system of conjugate harmonic functions can be identified with (the components of) a Clifford analytic function. Altogether, the results presented here can be regarded as a partial answer to the problem posed by Dahlberg in [Dah3] inquiring about the possibility of extending to higher dimensions the theory developed in $\mathbb{C}$ via conformal mapping techniques from [Ke].

As for Chapter 5, we submit that the classical boundary value problems for the Laplace operator in Lipschitz domains can be very naturally treated with the aid of Clifford algebra techniques. An abstraction of the main idea is that to any reasonable harmonic function one can append a "tail" so that the resulting function is Clifford analytic and has roughly the same "size" as the initial one. We also discuss several other applications, including a Burkholder-Gundy-Silverstein type result which is very close in spirit to the original theorem (cf. [BGS]).

Several exercises outline further developments and complement the body of results in each chapter.

I would like to express my sincere appreciation and gratitude to the people with whom I have discussed various aspects of this book during its elaboration. In particular, thanks are due to Björn Jawerth who actually suggested the writing of this book, for reading preliminary drafts and for his many constructive suggestions. Several enriching discussions with Alan McIntosh, Paul Koosis, Richard Delanghe and Margaret Murray are also acknowledged with gratitude.

Last but not least, I wish to thank Professor Martin Jurchescu for the trust, inspiration, guidance and moral support he generously (and constantly) gave me over the years.

Chapter 1
Clifford Algebras

This chapter is an overview of some basic facts concerning Clifford algebras (cf. also [BDS] and [GM2]; see also [MS] for some related historical clues). Here we set up the general formalism commonly used in the sequel.

§1.1 REAL AND COMPLEX CLIFFORD ALGEBRAS

Definition 1.1. *The Clifford algebra associated with $\mathbb{R}^n$, endowed with the usual Euclidean metric, is the extension of $\mathbb{R}^n$ to a unitary, associative algebra $\mathbb{R}_{(n)}$ over the reals, for which*

(1) $x^2 = -|x|^2$, *for any $x \in \mathbb{R}^n$;*

(2) $\mathbb{R}_{(n)}$ *is generated (as an algebra) by $\mathbb{R}^n$;*

(3) $\mathbb{R}_{(n)}$ *is not generated (as an algebra) by any proper subspace of $\mathbb{R}^n$.*

By polarization (1) becomes

$$xy + yx = -2\langle x, y \rangle, \tag{1.1}$$

for any $x, y \in \mathbb{R}^n$, where $\langle \cdot, \cdot \rangle$ stands for the usual inner product. In particular, if $\{e_j\}_{j=1}^n$ denotes the standard basis of $\mathbb{R}^n$, (1) is equivalent to

$$e_j e_k + e_k e_j = -2\,\delta_{jk}. \tag{1.2}$$

In other words $e_j^2 = -1$ for any $1 \le j \le n$, and $e_j e_k = -e_k e_j$ for any $1 \le j \ne k \le n$. Consequently, by (2), any element $a \in \mathbb{R}_{(n)}$ has a representation of the form

$$a = \sideset{}{'}\sum_I a_I e_I, \quad a_I \in \mathbb{R}. \tag{1.3}$$

Here $\sum'$ indicates that the sum is performed only over strictly increasing multi-indices I, i.e. ordered l-tuples of the form $I = (i_1, i_2, \ldots, i_l)$, with $1 \le i_1 < i_2 < \cdots < i_l \le n$,

1

where $0 \leq l \leq n$. Furthermore, e_I stands for the product $e_{i_1} \cdot e_{i_2} \cdot \ldots \cdot e_{i_l}$. By convention, $e_\varnothing := e_0 := 1$. We also set $e^I := (-1)^l e_{i_l} \cdot e_{i_{l-1}} \cdot \ldots \cdot e_{i_1}$, so that $e_I e^I = e^I e_I = 1$.

We shall see momentarily that the condition (3) is in fact equivalent to

$$e_{(1,2,\ldots,n)} \neq \pm 1,$$

hence, in even-dimensional cases it becomes superfluous. However, note that in general one must impose (3) to ensure the uniqueness of the Clifford algebra $\mathbb{R}_{(n)}$. This can be seen from the following simple example. One can embed $\mathbb{R}^3$ either in $\mathbb{H}$, the skew field of quaternions, by identifying e_1, e_2, e_3 with i, j, k, respectively, or in $\mathbb{H} \oplus \mathbb{H}$ this time by identifying e_1, e_2, e_3 with (i,j), (j,i), (k,k), respectively, and both embeddings satisfy (1) and (2) in Definition 1.1. Here i, j, k are the usual imaginary units in $\mathbb{H}$, i.e. $i^2 = j^2 = -1$, $ij = -ji = k$.

Theorem 1.2. *The Clifford algebra $\mathbb{R}_{(n)}$ exists and is uniquely determined up to an algebra isomorphism.*

Before proceeding with the proof of this result, we point out that the Clifford algebras $\mathbb{R}_{(0)}$, $\mathbb{R}_{(1)}$ and $\mathbb{R}_{(2)}$ are the real numbers, complex numbers, and quaternions, respectively.

Proof. Let us first indicate the proof of the uniqueness part. Obviously, it suffices to show that if a in (1.3) is zero, then necessarily $a_I = 0$ for all I. To this effect, we use the following identity (whose proof we postpone for the moment)

$$2^{-n} \sum_I{}' e_I \, a \, e^I = \begin{cases} a_\varnothing, & \text{if } n \text{ is even,} \\ a_\varnothing + a_{(1,2,\ldots,n)} e_{(1,2,\ldots,n)}, & \text{if } n \text{ is odd.} \end{cases} \tag{1.4}$$

The minimality of $\mathbb{R}_{(n)}$ ensures that 1 and $e_{(1,2,\ldots,n)}$ are linearly independent over the reals so that, at any rate, (1.4) implies that $a_\varnothing = 0$. Using the above reasoning with $e^I a$ in place of a, for arbitrary I, the conclusion follows.

As far as the existence of such an algebra is concerned, we shall produce an example in the matrix algebra $\mathcal{M}_{2^n \times 2^n}(\mathbb{R})$. Consider the matrices $e_j := E_j^n$, $j = 1, 2, \ldots, n$, where, for each $1 \leq k \leq n$, $\{E_j^k\}_{j=1}^k$ are inductively defined by

$$E_1^1 := \begin{pmatrix} 0 & -1 \\ 1 & 0 \end{pmatrix}$$

2

and, in general, for $1 \leq k \leq n-1$, $1 \leq j \leq k$,

$$E_j^{k+1} := \begin{pmatrix} E_j^k & 0 \\ 0 & -E_j^k \end{pmatrix}, \quad \text{and } E_{k+1}^{k+1} := \begin{pmatrix} 0 & -I_{2^k} \\ I_{2^k} & 0 \end{pmatrix}.$$

It is easy to see that $\{e_j\}_j$ satisfy (1.2). Also, since the trace of e_I is zero for all $I \neq \varnothing$, it is easy to see that $\{e_I\}_I$ are linearly independent over $\mathbb{R}$. Thus, we may take $\mathbb{R}_{(n)}$ to be the sub-algebra of $\mathcal{M}_{2^n \times 2^n}(\mathbb{R})$ consisting of all matrices a of the form (1.3). ∎

Proof of (1.4). Let $|\cdot|$ denote the cardinality function. Since for $I, J \subseteq \{1, 2, ..., n\}$, $e_I e_J = (-1)^{|I||J| - |I \cap J|} e_J e_I$, we have, for an arbitrary $x = \sum'_J x_J e_J$ in $\mathbb{R}_{(n)}$,

$$\sum_I{}' e_I x e^I = \sum_{I,J}{}' x_J e_I e_J e^I = \sum_{I,J}{}' (-1)^{|I||J| - |I \cap J|} x_J e_J.$$

Going further, for a fixed $J \subseteq \{1, 2, ..., n\}$,

$$\sum_I{}' (-1)^{|I||J| - |I \cap J|} = \sum_{i=0}^{|J|} \sum_{j=0}^{n-|J|} \sum_{\substack{|I|=i+j \\ |I \cap J|=i}}{}' (-1)^{(i+j)|J| - i}$$

$$= \sum_{i=0}^{|J|} \sum_{j=0}^{n-|J|} (-1)^{j|J|}(-1)^{i(|J|-1)} C_{|J|}^i C_{n-|J|}^j$$

$$= (1 + (-1)^{|J|})^{n-|J|}(1 + (-1)^{|J|-1})^{|J|} = 0,$$

unless either n is odd and $J = \{1, 2, ..., n\}$, or $J = \varnothing$. In these cases we end up with 2^n and, hence, (1.4) follows. ∎

Note that this proof actually gives more. First, $\dim \mathbb{R}_{(n)} = 2^n$ and we have a natural conjugation on $\mathbb{R}_{(n)}$, denoted by $\bar{}$, given by the usual transposition of matrices in $\mathcal{M}_{2^n \times 2^n}(\mathbb{R})$.

In the sequel, it will be also useful to embed $\mathbb{R}^{n+1}$ into $\mathbb{R}_{(n)}$ by identifying $(x_0, x) \in \mathbb{R}^{n+1} = \mathbb{R} \oplus \mathbb{R}^n$ with $x_0 \cdot e_0 + x \in \mathbb{R}_{(n)}$ (note that, by (1) in Definition 1.1, $1 \notin \mathbb{R}^n$), and call these elements *Clifford vectors*. An important observation is that any Clifford vector X has a multiplicative inverse, given by $X^{-1} = \overline{X}/|X|^2$. The multiplicative group generated by all Clifford vectors in $\mathbb{R}_{(n)}$ is called *the Clifford group*.

3

We define *the real part* of $a \in \mathbb{R}_{(n)}$ as Re $a := a_\varnothing$, if a is as in (1.3), and endow $\mathbb{R}_{(n)}$ with the natural Euclidean metric $|a|^2 := \text{Re}\,(a\bar{a}) = \text{Re}\,(\bar{a}a)$. It is not difficult to check that $|xy| \leq 2^{n-1}|x||y|$, for any $x, y \in \mathbb{R}_{(n)}$, and that $|xy| = |x||y|$ if at least one of x, y belongs to the Clifford group of $\mathbb{R}_{(n)}$. We also note that a similar construction works for the complex case, too. The resulting 2^n dimensional complex algebra will be denoted by $\mathbb{C}_{(n)}$.

In the last part of this section we shall discuss the exponential map

$$\exp x := \sum_{k=0}^\infty \frac{x^k}{k!}, \quad x \in \mathbb{R}_{(n)}.$$

In general, for $x \in \mathbb{R}_{(n)}$, $|\exp x| \leq \exp{(2^{n-1}|x|)}$, but if x is a Clifford number then this estimate becomes more precise: $|\exp x| = \exp{(\text{Re}\,x)}$. Our main result concerning this function is the next theorem.

Theorem 1.3. *exp* $: \mathbb{R}^{n+1} \longrightarrow \mathbb{R}^{n+1} \setminus \{0\}$ *is well-defined and onto.*

An important consequence is the existence of the N-th root for Clifford vectors.

Corollary 1.4. *For each integer $N \geq 1$ and each $u \in \mathbb{R}^{n+1}$, the equation $x^N = u$ has a solution in $\mathbb{R}^{n+1}$.*

Proof. If $u = 0$, we pick $x = 0$. Otherwise $u = \exp y$ for some $y \in \mathbb{R}^{n+1}$, and then we may take $x := \exp{(y/N)}$. ∎

Proof of Theorem 1.3. It is easy to see that the product of two Clifford vectors is again a Clifford vector if and only if they commute. Using this observation and proceeding inductively the well-definiteness of exp follows.

Let us introduce one more piece of notation. For arbitrary $u \in \mathbb{R}^{n+1}$, we set

$$\mathcal{A}(u) := \overline{\{p(u, \bar{u})\,;\, p \in \mathbb{R}[X, Y]\}},$$

i.e. $\mathcal{A}(u)$ is the smallest closed, commutative C^*-subalgebra of $\mathbb{R}_{(n)}$ which contains u. We now fix $u \in \mathbb{R}^{n+1}$, $u \neq 0$, and prove the existence of Clifford vector x so that $\exp x = u$. First, we claim that if u has Re $u > 0$, then there exists x in $\mathcal{A}(u)$ such that $\exp x = u$. To see this, we write $u = M(1 + (u - M)/M)$ for an arbitrary positive constant M, so that if $|u - M| < M^{-1}$ we can define

$$y := \sum_{k=1}^{\infty} \frac{1}{k} \left(\frac{u - M}{M} \right)^k \in \mathcal{A}(u).$$

Since $\exp y = u/M$, $x := y + \log M$ will do. Hence, proving the claim comes down to finding such a M. To this effect, it suffices to note that, as a function of M, $|(u - M)/M|^2 = |u|^2/M^2 - 2\mathrm{Re}\, u/M + 1$ attains its minimum $1 - (\mathrm{Re}\, u)^2/|u|^2 < 1$ at $M := |u|^2/(\mathrm{Re}\, u) > 0$.

Next, we treat the case $u_0 := \mathrm{Re}\, u \leq 0$. For any real number a we have that $(a + \overline{u})u = au + |u|^2 \in \mathbb{R}^{n+1}$, $\mathrm{Re}\,(au + |u|^2) = au_0 + |u|^2$ and $\mathrm{Re}\,(a + \overline{u}) = a + u_0$. We are looking for an a such that $au_0 + |u|^2 > 0$ and $a + u_0 > 0$. If $u_0 = 0$ any positive number will do, whereas for $u_0 < 0$ any a from the interval $(-u_0, -|u|^2/u_0)$ does the job. The only case when this interval degenerates is for $u = u_0$, but then e.g. $x := \log(-u_0) + \pi e_1$ solves the problem.

Taking a with these properties, the above claim then shows that we can find two Clifford vectors $x \in \mathcal{A}(au + |u|^2) = \mathcal{A}(u)$ with $\exp x = (a + \overline{u})u$, and $y \in \mathcal{A}(a + \overline{u}) = \mathcal{A}(u)$ with $\exp y = a + \overline{u}$. Since $x, y \in \mathcal{A}(u)$, it follows that x and y commute so that $u = (\exp x)(\exp y)^{-1} = \exp(x - y)$, and the proof is complete. ∎

Exercise. Use the identity (1.4) to describe the center of the Clifford algebra $\mathbb{R}_{(n)}$.

Before closing this section, let us introduce a notational convention which will be constantly used in the sequel. The estimate $F \lesssim G$, for two quantities depending on some parameter $s \in S$, signifies that there exists a positive constant C such that $F(s) \leq C\, G(s)$, for all $s \in S$. Furthermore, $F \approx G$ stands for $F \lesssim G$ and $G \lesssim F$.

§1.2 ELEMENTS OF CLIFFORD ANALYSIS

In this section we shall work in the Euclidean space $\mathbb{R}^{n+1}$, assumed to be embedded in the Clifford algebra $\mathbb{R}_{(n)}$. Let f, g be two locally bounded, Clifford valued functions defined in an open domain $\Omega \subseteq \mathbb{R}^{n+1}$. Inspired by Pompeiu's "dérivée aréolaire" ([Po2]), we introduce the Clifford derivative of the ordered pair (f, g) at a point $X \in \Omega$ by

$$D(f|g)(X) := \lim_{Q \downarrow X} \frac{\int_{\partial Q} f\, n\, g\, d\sigma}{\iint_Q d\,\mathrm{Vol}}.$$

More specifically, (f,g) is *Clifford differentiable* at X if there exists an element $c \in \mathbb{R}_{(n)}$ such that, for any $\epsilon > 0$, there exists an open neighborhood $U \subseteq \Omega$ of X so that

$$\left| \int_{\partial Q} f\, n\, g\, d\sigma - c\, \mathrm{Vol}\,(Q) \right| < \epsilon\, \mathrm{Vol}\,(Q),$$

for all rectangles Q of $\mathbb{R}^{n+1}$ with $X \in Q \subset U$ and such that, for some a priori fixed positive number C, $\mathrm{Vol}\,(Q) \geq C\,\mathrm{diam}\,(Q)^{n+1}$. Here n is the outward unit normal to the boundary of the rectangle Q, $d\sigma$ is the surface measure of ∂Q, and Vol stands for the usual Lebesgue measure in $\mathbb{R}^{n+1}$. Also, the top integrand must be interpreted in the sense of point-wise multiplication of Clifford algebra valued functions.

Let us call the ordered pair (f,g) *absolutely continuous on* Ω ([Ju]) if for any rectangle $Q \subset \Omega$ and any $\epsilon > 0$ there exists $\delta > 0$ such that

$$\sum_{i \in J} \left| \int_{\partial Q_i} f\, n\, g\, d\sigma \right| \leq \epsilon,$$

for any finite rectangular subdivision $(Q_i)_{i \in I}$ of Q, and any subset $J \subseteq I$ for which $\sum_{i \in J} \mathrm{Vol}\,(Q_i) \leq \delta$.

It is easy to check that if, for instance, both f and g are locally Lipschitz continuous then (f,g) is absolutely continuous. The importance of the notion of absolute continuity resides in the following.

Theorem 1.5. *If (f,g) is absolutely continuous on Ω, then $D(f|g)$ exists at almost any point of Ω. Moreover, $D(f|g)$ is locally integrable on Ω.*

Sketch of proof. ([Ju]) For any rectangle Q of $\mathbb{R}^{n+1}$ which is contained in Ω set

$$\rho(Q) := \sup \left\{ \sum_{i \in I} \left| \int_{\partial Q_i} f\, n\, g\, d\sigma \right| ; (Q_i)_{i \in I} \text{ finite rectangular subdivision of } Q \right\},$$
$$(1.5)$$

so that $\left| \int_{\partial Q} f n g\, d\sigma \right| \leq \rho(Q) < +\infty$ for any Q. Also, since $Q \mapsto \int_{\partial Q} f n g\, d\sigma$ is *rectangle-additive*, i.e. $\int_{\partial Q} f n g\, d\sigma = \sum_{i \in I} \int_{\partial Q_i} f n g\, d\sigma$ for any rectangle Q and any rectangular subdivision $(Q_i)_{i \in I}$ of Q, so is ρ. Next, we extend the action of ρ to the collection of all compact subsets of Ω by setting

$$\rho(K) := \inf \left\{ \sum_{i \in I} \rho(Q_i) ; K \subseteq \bigcup_{i \in I} Q_i \right\},$$

6

where the infimum is taken over all finite collection of rectangles $(Q_i)_{i \in I}$ included in Ω and having mutually disjoint interiors.

As ρ is rectangle-additive, this extension is consistent with the initial definition of ρ. Also, due to the absolute continuity of (f, g), ρ becomes *continuous* in the sense that $\rho(K_\nu) \longrightarrow \rho(K)$, whenever $\{K_\nu\}_\nu$ is a nested sequence of compacts in Ω such that $\cap_\nu K_\nu = K$.

For any multi-index $\alpha \in \mathbb{N}^{n+1}$ and for any $\nu \in \mathbb{N}$, we introduce $Q_{\nu,\alpha} := [0, 2^{-\nu}]^{n+1} + 2^{-\nu}\alpha$, and $I_\nu := \{\alpha \in \mathbb{N}^{n+1} \,; \, Q_{\nu,\alpha} \subseteq \Omega\}$. Also, for any real-valued, compactly supported function $\varphi \in C_0(\Omega)$, we set

$$I_\nu(\varphi) := \{\alpha \in I_\nu \,; \, \text{supp}\, \varphi \cap Q_{\nu,\alpha} \neq \varnothing\}$$

and

$$P_\nu(\varphi) := \bigcup_{\alpha \in I_\nu(\varphi)} Q_{\nu,\alpha}.$$

It follows that $P_{\nu+1}(\varphi) \subseteq P_\nu(\varphi)$ for any ν and $\cap_\nu P_\nu(\varphi) = \text{supp}\, \varphi$. If we now introduce

$$s_\nu(\varphi) := \sum_{\alpha \in I_\nu(\varphi)} \varphi(2^{-\nu}\alpha) \int_{\partial Q_{\nu,\alpha}} f \, n \, g \, d\sigma,$$

then s_ν is $\mathbb{R}$–linear and satisfies

$$|s_\nu(\varphi)| \leq \rho(P_\nu(\varphi))\|\varphi\|_\infty, \quad \varphi \in C_0(\Omega).$$

Finally, we define $\mu : C_0(\Omega) \longrightarrow \mathbb{R}_{(n)}$ by setting

$$\mu(\varphi) := \lim_\nu s_\nu(\varphi),$$

where the existence of the limit easily follows from the uniform continuity of φ. Since μ is $\mathbb{R}$–linear and satisfies $|\mu(\varphi)| \leq \rho(\text{supp}\, \varphi)\|\varphi\|_\infty$, we conclude that μ is a Clifford algebra valued Radon measure on Ω.

Next, we fix an arbitrary rectangle $Q \subset \Omega$ and take $\varphi_\nu \in C_0(\Omega)$ a sequence of real-valued functions such that $0 \leq \varphi_\nu \leq 1$ on Ω, $\varphi_\nu = 1$ in a neighborhood of Q, $\text{supp}\, \varphi_{\nu+1} \subseteq \text{supp}\, \varphi_\nu$ and $\cap_\nu \text{supp}\, \varphi_\nu = Q$. From the definition of μ,

$$\left| \mu(\varphi_\nu) - \int_{\partial Q} f \, n \, g \, d\sigma \right| \leq \rho(\text{supp}\, \varphi_\nu) - \rho(Q),$$

7

hence, by the continuity of ρ, $\int_Q d\mu = \int_{\partial Q} fng\,d\sigma$, for any Q. Using this and once again the absolute continuity of (f,g), we infer that μ is absolutely continuous with respect to the Lebesgue measure. Therefore, if $h \in L^1(\Omega, \text{loc})$ denotes the Radon-Nikodym-Lebesgue density of μ with respect to the Lebesgue measure, we have that

$$\int_{\partial Q} f\,n\,g\,d\sigma = \int_Q d\mu = \iint_Q h.$$

Using this and Lebesgue's differentiation theorem we finally obtain that $D(f|g) = h$, and this completes the proof of the theorem. ∎

The next lemma, which in the complex case goes back to Pompeiu [Po3], (cf. also [Te], [JM]), can be though of as the higher dimensional analogue of the classical Leibnitz-Newton formula on the real line. Recall that a bounded Lipschitz domain in $\mathbb{R}^{n+1}$ is a bounded domain whose boundary is locally given by the graph of a Lipschitz function (see e.g. [Gr], [Ne]).

Lemma 1.6. *Let Ω be a bounded Lipschitz domain in $\mathbb{R}^{n+1}$ and let f, g be two Clifford algebra valued, continuous functions on $\bar{\Omega}$ such that $D(f|g)$ is also continuous on $\bar{\Omega}$. Then*

$$\int_{\partial\Omega} f\,n\,g\,d\sigma = \iint_\Omega D(f|g)\,d\text{Vol}. \tag{1.6}$$

Sketch of Proof. We shall use a Goursat type argument. Reasoning by contradiction and assuming $\left| \int_{\partial\Omega} f\,n\,g\,d\sigma - \iint_\Omega D(f|g) \right| \geq c > 0$, an usual partitioning argument yields a sequence of nested domains $(\omega_j)_j$, with $\bigcap_j \omega_j = \{X_0\}$, for some $X_0 \in \bar{\Omega}$, and $\text{Vol}(\omega_j) \approx 2^{-j(n+1)}\text{Vol}(\omega)$, such that

$$\left| \int_{\partial\omega_j} f\,n\,g\,d\sigma - \iint_{\omega_j} D(f|g) \right| \geq 2^{-j(n+1)}c.$$

Dividing by $\text{Vol}(\omega_j)$ and using the fact that $\frac{1}{\text{Vol}(\omega_j)} \int_{\partial\omega_j} f\,n\,g\,d\sigma \longrightarrow D(f|g)(X_0)$ by definition, whereas $\frac{1}{\text{Vol}(\omega_j)} \iint_{\omega_j} D(f|g) \longrightarrow D(f|g)(X_0)$ by the continuity of $D(f|g)$, we finally contradict the original assumption. ∎

If f, g are Lipschitz continuous, say, it is easy to check the Leibnitz rule $D(f|g) = D(f|1)\,g + f\,D(1|g)$, and we shall simply set $Dg := D(1|g)$ and $fD := D(f|1)$. Note that Lemma 1.6 gives

$$\int_{\partial\Omega} f\,n\,g\,d\sigma = \iint_\Omega \{(fD)g + f(Dg)\}\,d\text{Vol}. \tag{1.7}$$

8

We also set $\overline{D}f := \overline{D(\overline{f}|1)}$ and $f\overline{D} := \overline{D(1|\overline{f})}$. It should be pointed out that at any point of differentiability $X \in \Omega$ of $f = \sum'_I f_I e_I$, we have

$$(Df)(X) = \sum_I' \sum_{j=0}^n \frac{\partial f_I}{\partial x_j}(X)e_j e_I,$$

and

$$(fD)(X) = \sum_I' \sum_{j=0}^n \frac{\partial f_I}{\partial x_j}(X)e_I e_j.$$

The verification is straightforward. Note that, by linearity considerations, it actually suffices to treat the case of a scalar valued function f. We can also assume that the point of differentiability is the origin of the system. In this later case, expanding f into its first order Taylor series around the origin

$$f(X) = f(0) + \sum_j x_j(\partial_j f)(0) + o(|X|), \quad X = (x_j)_j \in \mathbb{R}^{n+1},$$

and using the easily checked fact that $\int_{\partial Q} x_j n \, d\sigma = e_j \mathrm{Vol}\,(Q)$, for any j, the conclusion follows.

Going further, simple calculations give that the Laplace operator $\triangle$ in $\mathbb{R}^{n+1}$ has the factorizations

$$\triangle = D\overline{D} = \overline{D}D. \tag{1.8}$$

Following Moisil and Teodorescu [MT], we shall call f *left monogenic* (*right monogenic*, or *two-sided monogenic*, respectively) if $Df = 0$ ($fD = 0$, or $Df = fD = 0$, respectively). Note that, by (1.8), any monogenic function is harmonic.

Our basic example of a two-sided monogenic function, the so called *Cauchy kernel*, is the fundamental solution of the operator D

$$E(X) := \frac{1}{\sigma_n} \frac{\overline{X}}{|X|^{n+1}}, \quad X \in \mathbb{R}^{n+1} \setminus \{0\}, \tag{1.9}$$

where σ_n stands for the area of the unit sphere in $\mathbb{R}^{n+1}$. This can be readily seen from (1.8) and $E = \overline{D}\Gamma_{n+1} = \Gamma_{n+1}\overline{D}$, where

$$\Gamma_{n+1}(X) := \begin{cases} \dfrac{1}{(1-n)\sigma_n} \dfrac{1}{|X|^{n-1}}, & X \neq 0, \quad n \geq 2, \\ \dfrac{1}{2\pi} \log|X|, & X \neq 0, \quad n = 1, \end{cases}$$

9

is the canonical fundamental solution for the Laplacean in $\mathbb{R}^{n+1}$. In fact, our next result shows that any left (or right) monogenic function which is $\mathbb{R}^{n+1}$-valued is necessarily two-sided monogenic.

Proposition 1.7. *Let $F = u_0 - \sum_{j=1}^{n} u_j e_j$ be a $\mathbb{R}^{n+1}$-valued function defined on a open set Ω of $\mathbb{R}^{n+1}$. The following are equivalent:*

(1) *The $(n+1)$-tuple $U := (u_j)_{j=0}^{n}$ is a system of conjugate harmonic functions in Ω in the sense of Moisil-Teodorescu [Mo3], [MT] and Stein-Weiss [SW1], i.e. it satisfies the so called generalized Cauchy-Riemann equations $\operatorname{div} U = 0$ and $\operatorname{curl} U = 0$ in Ω;*

(2) *F is left monogenic in Ω;*

(3) *F is right monogenic in Ω;*

(4) *The 1-form $\omega := u_0 dx_0 - u_1 dx_1 - \ldots - u_n dx_n$ has $d\omega = 0$ and $d^*\omega = 0$ in Ω, where d and d^* are the exterior differentiation operator and its formal transpose, respectively.*

In addition, if the domain Ω is simply connected, then the above conditions are further equivalent to

(5) *There exists a unique (modulo an additive constant) real valued harmonic function U in Ω such that $(u_j)_{j=0}^{n} = \operatorname{grad} U$ in Ω (i.e. $F = \overline{D}U$).*

The easy proof is omitted.

Lemma 1.6 applied to f and $g := E(X - \cdot)$ in $\Omega \setminus B_\epsilon(X)$ yields, after letting ϵ go to zero, the Clifford version of Pompeiu's integral representation formula ([Po1], [Mo1,2], [Te]).

Therem 1.8. *Let Ω be a bounded Lipschitz domain in $\mathbb{R}^{n+1}$. If f and Df are continuous on $\bar{\Omega}$, then*

$$f(X) = \mathcal{C}f(X) + T(Df)(X), \quad X \in \Omega,$$

where

$$\mathcal{C}f(X) := \frac{1}{\sigma_n} \int_{\partial\Omega} \frac{\overline{Y - X}}{|Y - X|^{n+1}} n(Y) f(Y) \, d\sigma(Y), \quad X \in \Omega,$$

and

$$Tf(X) := \frac{1}{\sigma_n} \iint_{\Omega} \frac{\overline{X - Y}}{|X - Y|^{n+1}} f(Y) \, d\operatorname{Vol}(Y), \quad X \in \Omega.$$

10

A similar formula for the left action of D holds as well.

As a corollary, let us note the Cauchy type reproducing formulas ([**Di**], [**MT**])

$$f(X) = \frac{1}{\sigma_n} \int_{\partial\Omega} \frac{\overline{Y-X}}{|Y-X|^{n+1}} n(Y) f(Y) \, d\sigma(Y), \quad X \in \Omega, \qquad (1.10)$$

if f is left monogenic in Ω, and

$$f(X) = \frac{1}{\sigma_n} \int_{\partial\Omega} f(Y) \, n(Y) \frac{\overline{Y-X}}{|Y-X|^{n+1}} \, d\sigma(Y), \quad X \in \Omega, \qquad (1.11)$$

if f is right monogenic in Ω.

For f right monogenic and g left monogenic in Ω, we also obtain from (1.7) the Cauchy type vanishing formula

$$\int_{\partial\Omega} f(X) \, n(X) \, g(X) \, d\sigma(X) = 0. \qquad (1.12)$$

Exercise. Let Ω be a bounded domain with C^∞ boundary.

• Prove that C maps $C^\infty(\partial\Omega)$ into $C^\infty(\overline{\Omega})$ and that T maps $C^\infty(\overline{\Omega})$ into itself.

• Show that $D(Tf) = f$ on $C^\infty(\overline{\Omega})$.

• Use this and the identity (1.8) to solve the Poisson equation $\triangle u = v$ in Ω, for arbitrary real-valued data $v \in C^\infty(\overline{\Omega})$.

§1.3 CLIFFORD MODULES

The "Clifordized" version $V_{(n)}$ of an arbitrary complex vector space V is defined by

$$V_{(n)} := V \otimes \mathbb{C}_{(n)} = \left\{ x = {\sum_I}' x_I \otimes e_I \, ; \, x_I \in V \text{ for all } I \right\}.$$

Thus $V_{(n)}$ becomes a two-sided Clifford module (that is, a two-sided module over the ring $\mathbb{C}_{(n)}$), by setting

$$\alpha x := {\sum_{I,J}}' \alpha_J x_I \otimes e_J e_I, \quad x\alpha := {\sum_{I,J}}' \alpha_J x_I \otimes e_I e_J,$$

11

for $x = \sum'_I x_I \otimes e_I \in V_{(n)}$ and $\alpha = \sum'_J \alpha_J e_J \in \mathbb{C}_{(n)}$. Moreover, if $(V, \|\cdot\|)$ is a normed vector space, then we endow $V_{(n)}$ with the Euclidean norm

$$\left\| \sum'_I x_I \otimes e_I \right\|_{(n)} := \left(\sum'_I \|x_I\|^2 \right)^{1/2}.$$

If $W \subseteq V_{(n)}$ is a left-(or right-)submodule of $V_{(n)}$, then any morphism of Clifford modules $L : W \longrightarrow \mathbb{C}_{(n)}$ is called a *Clifford functional* of W. The collection of all Clifford functionals of W will be denoted by W^*.

Consider now $\mathcal{H}$ a complex Hilbert space (fixed for the rest of this section) and let $(\cdot, \cdot)$ be the corresponding inner product on $\mathcal{H}$. Then $\mathcal{H}_{(n)}$ becomes a complex Hilbert space when endowed with

$$[x, x] := \|x\|_{(n)}^2 = \sum'_I \|x_I\|^2 := \sum'_I (x_I, x_I),$$

if $x = \sum'_I x_I \otimes e_I \in \mathcal{H}_{(n)}$. We also introduce the following $\mathbb{C}_{(n)}$−valued form on $\mathcal{H}_{(n)}$

$$\langle x, y \rangle := \sum'_{I,J} (x_I, y_J)\, e_I \overline{e_J},$$

if $x = \sum'_I x_I \otimes e_I \in \mathcal{H}_{(n)}$, $y = \sum'_J y_J \otimes e_J \in \mathcal{H}_{(n)}$. Finally, suppose that $\mathcal{H}$ has an involutive structure, i.e. there exists a conjugate linear isometry of $\mathcal{H}$, denoted by $\bar{\cdot}$, such that $\bar{\bar{x}} = x$ for any x in $\mathcal{H}$. We extend this involution to $\mathcal{H}_{(n)}$ by introducing

$$\bar{x} := \sum'_I \overline{x_I} \otimes \overline{e_I}, \quad \text{if } x = \sum'_I x_I \otimes e_I \in \mathcal{H}_{(n)}.$$

Call an element $x \in \mathcal{H}_{(n)}$ *self-adjoint* provided $\bar{x} = x$. Similarly, a two-sided submodule V of $\mathcal{H}_{(n)}$ is said to be *self-adjoint* if $\overline{V} := \{\bar{x};\ x \in V\} = V$.

The main, elementary properties of these objects are collected in the next proposition.

Proposition 1.9. *For x, $y \in \mathcal{H}_{(n)}$ and $\alpha \in \mathbb{C}_{(n)}$ the following hold.*

(1) $\langle x, y \rangle = \sum'_I [\overline{e_I}x, y] e_I$. *In particular,* $\mathrm{Re}\,\langle x, y \rangle = [x, y]$.

(2) $\mathrm{Re}\,\langle x, x \rangle = \|x\|_{(n)}^2$. *In particular,* $\|x\|_{(n)}^2 \approx |\langle x, x \rangle|$.

12

(3) $|\langle x, y \rangle| \le \|x\|_{(n)} \|y\|_{(n)}$.

(4) *If $\lambda \in \mathbb{R}^{n+1} \subset \mathbb{C}_{(n)}$, then $\|\lambda x\|_{(n)} = |\lambda| \|x\|_{(n)}$.*

   In particular $\|\lambda x\|_{(n)} \le 2^{n-1} |\lambda| \|x\|_{(n)}$ if $\lambda \in \mathbb{R}_{(n)}$ and, in general, if $\alpha \in \mathbb{C}_{(n)}$, $\|\alpha x\|_{(n)} \le 2^{n-1/2} |\alpha| \|x\|_{(n)}$.

(5) $\langle \alpha x, y \rangle = \alpha \langle x, y \rangle, \quad \langle x, \alpha y \rangle = \langle x, y \rangle \overline{\alpha}$.

   Also, $\langle x\alpha, y \rangle = \langle x, y\overline{\alpha} \rangle$ and $\overline{\langle x, y \rangle} = \langle y, x \rangle$.

(6) $\overline{\overline{x}} = x, \quad \overline{\alpha \overline{x}} = \overline{x}\,\overline{\alpha}, \quad \overline{x\alpha} = \overline{\alpha}\,\overline{x}, \quad (\overline{x}, \overline{y}) = \overline{(x, y)}$.

Proof. Let us consider (4), for example. We have

$$\|x\lambda\|_{(n)}^2 = \mathrm{Re}\,\langle x\lambda, x\lambda \rangle = \mathrm{Re}\,\langle x\lambda\overline{\lambda}, x \rangle = |\lambda|^2 \mathrm{Re}\,\langle x, x \rangle = |\lambda|^2 \|x\|_{(n)}^2.$$

Thus, $\|\lambda x\|_{(n)} = \|\overline{\lambda x}\|_{(n)} = \|\overline{x}\,\overline{\lambda}\|_{(n)} = |\overline{\lambda}| \|\overline{x}\|_{(n)} = |\lambda| \|x\|_{(n)}$. ∎

Remark. The notion of monogenicity has a natural extension in the context of $\mathcal{H}_{(n)}$−valued functions. More specifically, $F : \Omega \longrightarrow \mathcal{H}_{(n)}$, where Ω is an open set in $\mathbb{R}^{n+1}$, is called left monogenic provided $\langle F(\cdot), h \rangle$ is left monogenic in the usual sense in Ω, for any h in $\mathcal{H}_{(n)}$. Similarly, F is called right monogenic provided $\langle h, \overline{F(\cdot)} \rangle$ is right monogenic in Ω for any $h \in \mathcal{H}_{(n)}$. Adopting this convention, one can readily see that the results of the previous section continue to hold for $\mathcal{H}_{(n)}$−valued functions as well.

The next proposition is the analog of the usual Riesz representation theorem.

Proposition 1.10. *Let $\mathcal{H}$ be as before and let V be a closed left-(or right-)submodule of $\mathcal{H}_{(n)}$. Then for any $L \in V^*$ there exists a unique element $a \in V$ such that $L(x) = \langle x, a \rangle$ (or $L(x) = \langle a, \overline{x} \rangle$, respectively) for all $x \in V$. Moreover, $\|a\|_{(n)} \approx \|L\|$.*

Proof. The usual form of the Riesz representation theorem yields the existence of an a in V such that $\mathrm{Re}\,L(\cdot) = [\,\cdot\,, a] = \mathrm{Re}\,\langle \cdot\,, a \rangle$. Then, for any $x \in V$,

$$L(x) = \sum_I{}' \mathrm{Re}\,(\overline{e_I} L(x)) e_I = \sum_I{}' \mathrm{Re}\, L(\overline{e_I}\,x) e_I = \sum_I{}' \mathrm{Re}\,\langle \overline{e_I}\,x, a \rangle a_I$$

$$= \sum_I{}' \mathrm{Re}\,(\overline{e_I}\langle x, a \rangle) e_I = \langle x, a \rangle.$$

As for the uniqueness part, we simply remark that $\langle x, a \rangle = 0$ for any $x \in V$ implies $\|a\|_{(n)}^2 = \mathrm{Re}\,\langle a, a \rangle = 0$, i.e. $a = 0$. Finally, from the properties of $\langle \cdot, \cdot \rangle$ given in Proposition 1.9, the norm of L is easily seen to be equivalent with $\|a\|_{(n)}$. ∎

13

Corollary 1.11. *If V is a closed left-submodule, say, of $\mathcal{H}_{(n)}$ and*

$$\mathcal{B} : V \times V \longrightarrow \mathbb{C}_{(n)}$$

is a continuous form, additive in each variable and such that $\mathcal{B}(\alpha x, y) = \alpha \mathcal{B}(x, y)$ and $\mathcal{B}(x, \alpha y) = \mathcal{B}(x, y)\overline{\alpha}$, for $\alpha \in \mathbb{C}_{(n)}$, $x, y \in V$, then there exists a unique continuous endomorphism T of V so that

$$\mathcal{B}(x, y) = \langle Tx, y \rangle, \quad x, y \in V. \tag{1.13}$$

In what follows, $\mathcal{B}$ will be referred to as a continuous $\mathbb{C}_{(n)}$-*sesquilinear form* on V. $\mathcal{B}$ is called *non-degenerate* if for any $x \in V$

$$\mathcal{B}(x, y) = 0 \; \forall y \in V \iff x = 0, \text{ and } \mathcal{B}(y, x) = 0 \; \forall y \in V \iff x = 0.$$

Note that for a continuous $\mathbb{C}_{(n)}$-sesquilinear form $\mathcal{B}$ on V, the operator T given by (1.13) is an automorphism of V if and only if $\mathcal{B}$ is non-degenerate. Also, we call $\mathcal{B}$ a δ-*non-degenerate* form on V, provided

$$\sup_{\substack{y \in V \\ \|y\|_{(n)} \leq 1}} |\mathcal{B}(x, y)| \geq \delta \|x\|_{(n)}, \quad \text{and} \quad \sup_{\substack{y \in V \\ \|y\|_{(n)} \leq 1}} |\mathcal{B}(y, x)| \geq \delta \|x\|_{(n)},$$

for any $x \in V$. Actually, any non-degenerate form is δ-non-degenerate for some $\delta > 0$. Finally, call $\mathcal{B}$ *symmetric* if $\overline{\mathcal{B}(x, y)} = \mathcal{B}(y, x)$, for any $x, y \in V$.

Corollary 1.12. *If V is a closed left-submodule of $\mathcal{H}_{(n)}$ and $\mathcal{B}_i$, $i = 1, 2$, two continuous, non-degenerate $\mathbb{C}_{(n)}$-sesquilinear forms on V, then there exists a unique continuous automorphism T of V such that*

$$\mathcal{B}_1(Tx, y) = \mathcal{B}_2(x, y), \quad \text{for all } x, y \in V.$$

Proof. The results discussed above ensure the existence of two continuous automorphisms S_i of V for which $\mathcal{B}_i(x, y) = \langle S_i x, y \rangle$, $i = 1, 2$. Take $T := S_1^{-1} S_2$. ∎

Exercise. Let V be a normed complex vector space, and let X be a left-submodule of $V_{(n)}$. Then any continuous Clifford functional φ of X extends to a continuous one $\varphi : V_{(n)} \longrightarrow \mathbb{C}_{(n)}$, having comparable norm with the initial functional.

Hint: Use the classical version of the Hahn-Banach theorem to extend first the real part of φ as a continuous morphism of complex vector spaces $\operatorname{Re} \varphi : V_{(n)} \longrightarrow \mathbb{C}_{(n)}$, then check that $\varphi = \sum'_I \operatorname{Re} \varphi(\overline{e_I} \cdot) e_I$ is in fact a morphism of Clifford modules.

Exercise. Prove that $(X_{(n)})^* \simeq (X^*)_{(n)}$.

Consider $\pi_{(n)} : X_{(n)} \longrightarrow X_{(n)}$ defined for any $x = \sum'_I x_I \otimes e_I \in X_{(n)}$ by

$$\pi_{(n)} x := \begin{cases} x_\varnothing \otimes e_0, & \text{if } n \text{ is even,} \\ x_\varnothing \otimes e_0 + x_{\{1,2,\dots,n\}} \otimes e_{\{1,2,\dots,n\}}, & \text{if } n \text{ is odd.} \end{cases}$$

Also, for $S \subseteq X_{(n)}$, let $\langle S \rangle$ be the smallest two-sided submodule of $X_{(n)}$ containing S.

Exercise. If Y is a two-sided submodule of $X_{(n)}$, then $Y = \langle \pi_{(n)} Y \rangle$. In particular, if n is even, then any two-sided Clifford submodule of $X_{(n)}$ is of the form $Y_{(n)}$ for some linear subspace $Y \subseteq X$.

Chapter 2

Constructions of Clifford Wavelets

The aim of this chapter is to present constructions of systems of Clifford algebra-valued wavelet-like bases adapted to a Clifford algebra valued measure $b(x)\,dx$ in $\mathbb{R}^m$, where $b : \mathbb{R}^m \longrightarrow \mathbb{R}^{n+1} \subset \mathbb{C}_{(n)}$ is an essentially bounded function having intergal means bounded away from zero (e.g. $\operatorname{Re} b(x) \geq \delta > 0$ will do).

Because the complex Clifford algebra $\mathbb{C}_{(n)}$ is non-commutative, a distinguished feature of such a system is that it should be in fact a system of *pairs* of Clifford algebra valued functions, say $\{\Theta^L_{j,k}\}_{j,k}$ and $\{\Theta^R_{j,k}\}_{j,k}$, called *Clifford wavelets*. These Clifford wavelets must have some adequate smoothness, the cancellation properties

$$\langle \Theta^L_{j,k}, \Theta^R_{j',k'}\rangle_b = \delta_{j,j'}\delta_{k,k'},$$

and, also, form a Riesz frame for L^2, i.e.

$$f = \sum \langle f, \Theta^R_{j,k}\rangle_b \Theta^L_{j,k} = \sum \Theta^R_{j,k}\langle \Theta^L_{j,k}, f\rangle_b,$$

$$\|f\|^2 \approx \sum |\langle f, \Theta^R_{j,k}\rangle_b|^2 \approx \sum |\langle \Theta^L_{j,k}, f\rangle_b|^2,$$

for any L^2−integrable, Clifford algebra valued function f. Here the pairing $\langle\cdot,\cdot\rangle_b$ is defined by

$$\langle f_1, f_2\rangle_b := \int_{\mathbb{R}^m} f_1(x)b(x)f_2(x)\,dx.$$

In the first part of this chapter, §2.1–§2.4, we shall closely follow Meyer and Tchamitchian ([Me], [Tc]) and prove the existence of such systems of Clifford wavelets satisfying additional smoothness and decay properties:

$$\Theta^L_{j,k},\ \Theta^R_{j,k} \in \mathcal{C}^r(\mathbb{R}^m)_{(n)},$$

for an arbitrary, a priori fixed, nonnegative integer r, and

$$|\partial^\alpha \Theta^L_{j,k}(x)| + |\partial^\alpha \Theta^R_{j,k}(x)| \lesssim 2^{k(m/2+|\alpha|)}\exp\left(-\kappa|2^k x - j|\right), \quad \forall |\alpha| \leq r.$$

for some $\kappa > 0$.

Of course, one cannot hope to obtain these functions by the usual dilation-translation operations performed on some initial Θ, but the above estimates suggest that everything happens as if this is possible.

Finally, in the last part (§2.5), a generalization of the classical Haar wavelet system to the Clifford-algebra framework is presented (cf. [**CJS**]). Applications will be discussed in the following chapters.

§2.1 ACCRETIVE FORMS AND ACCRETIVE OPERATORS

Let $\mathcal{H}$ be an arbitrary complex separable Hilbert space, fixed throughout this chapter, and let V be a closed left-(or right-)Clifford submodule of $\mathcal{H}_{(n)}$. Following Kato [**Ka**], we call a continuous endomorphism T of V $\delta-accretive$ on V if

$$\mathrm{Re}\,[Tx, x] \geq \delta \|x\|_{(n)}^2, \quad x \in V.$$

Obviously, if T is δ-accretive, then T is an isomorphism of V. Also, we call a form

$$B : \mathcal{H}_{(n)} \times \mathcal{H}_{(n)} \longrightarrow \mathbb{C}_{(n)}$$

δ-$accretive$ on $\mathcal{H}_{(n)}$ if the following conditions are fulfilled.

(1) $B(\cdot, \cdot)$ is *Clifford bilinear*, i.e. $B(\alpha x + \beta y, z) = \alpha B(x, z) + \beta B(y, z)$, and $B(x, y\alpha + z\beta) = B(x, y)\alpha + B(x, z)\beta$ for all $\alpha, \beta \in \mathbb{C}_{(n)}$ and $x, y, z \in \mathcal{H}_{(n)}$.

(2) $B(\cdot, \cdot)$ is continuous on V, i.e. $|B(x, y)| \lesssim \|x\|_{(n)} \|y\|_{(n)}$, uniformly for $x, y \in \mathcal{H}_{(n)}$.

(3) $\mathrm{Re}\,B(x, \bar{x}) \geq \delta \|x\|_{(n)}^2$, for any $x \in \mathcal{H}_{(n)}$.

The next proposition provides us with a basic example in this respect.

Proposition 2.1. *If $b \in \mathbb{R}^{n+1} \hookrightarrow \mathbb{C}_{(n)}$ has $\mathrm{Re}\,b > 0$, then the form*

$$B(x, y) := \langle xb, \bar{y} \rangle, \quad x, y \in \mathcal{H}_{(n)},$$

is $(\mathrm{Re}\,b)$-accretive.

Proof. The only non-obvious property is (3). But for an arbitrary $x \in V$,

$$
\begin{aligned}
\operatorname{Re} B(x, \overline{x}) = \operatorname{Re} \langle xb, x \rangle &= \frac{1}{2} \operatorname{Re} \left\{ \langle xb, x \rangle + \overline{\langle xb, x \rangle} \right\} \\
&= \frac{1}{2} \operatorname{Re} \left\{ \langle xb, x \rangle + \langle x, xb \rangle \right\} = \frac{1}{2} \operatorname{Re} \left\{ \langle xb, x \rangle + \langle x\overline{b}, x \rangle \right\} \\
&= \frac{1}{2} \operatorname{Re} \langle x(b + \overline{b}), x \rangle = \operatorname{Re} b \operatorname{Re} \langle x, x \rangle = \operatorname{Re} b \| x \|_{(n)}^2 ,
\end{aligned}
$$

and the conclusion follows. ∎

Finally, note that for any δ-accretive form $B(\cdot, \cdot)$ on $\mathcal{H}_{(n)}$, $B(\cdot, \overline{\cdot})$ is a δ-non-degenerate form on $\mathcal{H}_{(n)}$.

§2.2 CLIFFORD MULTIRESOLUTION ANALYSIS. THE ABSTRACT SETTING

Our first result is the core of the algorithm we shall set up in the next section.

Proposition 2.2. *Let $V \subseteq H \subseteq \mathcal{H}_{(n)}$ be closed, two-sided submodules of $\mathcal{H}_{(n)}$, let B be a δ-non-degenerate form on H, and consider*

$$
X^L := \{ x \in H \, ; \, B(x, y) = 0 \text{ for all } y \in V \},
$$
$$
X^R := \{ x \in H \, ; \, B(y, x) = 0 \text{ for all } y \in V \}.
$$

Then :

(1) X^L *is a closed-left submodule of* $\mathcal{H}_{(n)}$, X^R *is a closed right-submodule in* $\mathcal{H}_{(n)}$ *and* $X^L \oplus V = V \oplus X^R = H$ *(non-orthogonal sums).*

(2) *The oblique projection operators from H parallel to V onto X^L and X^R, respectively, denoted by π^L and π^R, respectively, are continuous morphisms with operator norms bounded by a constant depending solely on n, $\|B\|$ and δ.*

(3) *Letting*

$$
W := \{ x \in H \, ; \, \langle x, y \rangle = 0 \text{ for all } y \in V \},
$$

then $W = H \ominus V$ (orthogonal difference) and, consequently, π^L, π^R project W isomorphically onto X^L and X^R, respectively. In addition, the operator norms of π^L and π^R are bounded from below by a constant depending only on n.

18

Proof. From the non-degeneracy of B we immediately get $X^L \oplus V \subseteq H$. By "Riesz lemma" (Proposition 1.10), there exist two left-Clifford-linear continuous operators $S : H \longrightarrow V$ and $T : V \longrightarrow V$ such that $B(h,v) = \langle Sh, \overline{v} \rangle$, for any $h \in H$, $v \in V$, and $B(u,v) = \langle Tu, \overline{v} \rangle$, for any $u, v \in V$. Also, the non-degeneracy of B implies that T is actually an automorphism of V. Let h be an arbitrary element in H and set $v := T^{-1}Sh \in V$. For any $u \in V$ we have

$$B(v,u) = \langle Tv, \overline{u} \rangle = \langle Sh, \overline{u} \rangle = B(h,u),$$

proving that in fact $X^L \oplus V = H$. Likewise, $V \oplus X^R = H$. Moreover, $\pi^L = I - T^{-1}S$ and $\|S\| \lesssim \|B\|$, $\|T^{-1}\| \leq \delta^{-1}$, so that (2) follows.

As for (3), simply note that on one hand $\operatorname{Ker} \pi^L = V$ and $V \cap W = \{0\}$, while on the other hand any $x \in X^L$ has a decomposition $x = v \oplus w \in V \oplus W$ and therefore

$$x = \pi^L x = \pi^L v + \pi^L w = \pi^L w.$$

Thus $\pi^L|_W$ is an isomorphism. In addition, if $x \in W$, then $x - \pi^L x$ belongs to V, hence $\langle x, x - \pi^L x \rangle = 0$. Consequently,

$$\|x\|_{(n)}^2 \leq |\langle x, x \rangle| = |\langle x, \pi^L x \rangle| \lesssim \|x\|_{(n)} \|\pi^L x\|_{(n)},$$

i.e. $\|x\|_{(n)} \lesssim \|\pi^L x\|_{(n)}$. ∎

We also note that if V and H in the above proposition are self-adjoint, then by (6) in Proposition 1.9, W is self-adjoint too.

We now make the following definition.

Definition 2.3. *A Clifford multiresolution analysis of $\mathcal{H}_{(n)}$ (CMRA for short) is any increasing sequence $\{V_k\}_{-\infty}^{+\infty}$ of closed, two-sided submodules of $\mathcal{H}_{(n)}$ such that*

$$\cap_{-\infty}^{+\infty} V_k = \{0\}, \quad \text{and} \quad \cup_{-\infty}^{+\infty} V_k \text{ is dense in } \mathcal{H}_{(n)},$$

together with a continuous Clifford bilinear form B on $\mathcal{H}_{(n)}$ such that, for some $\delta > 0$, $B(\cdot, \overline{\cdot})\big|_{V_k \times V_k}$ is a δ-non-degenerate form on V_k, for any $k \in \mathbb{Z}$.

19

Also, if for each k, X_k^L, X_k^R, W_k and π_k^L, π_k^R are as in Proposition 2.2 when one takes $V := V_{k-1}$ and $H := V_k$, then $\{X_k^L\}_k$, $\{X_k^R\}_k$ will be called the wavelet spaces of this CMRA.

Obviously, $\bigcup\limits_k V_k$ and $\oplus W_k$ have the same closure in $\mathcal{H}_{(n)}$, therefore $\oplus\limits_k W_k$ is dense in $\mathcal{H}_{(n)}$, and the next proposition tells us that something similar happens for the wavelet spaces of a CMRA of $\mathcal{H}_{(n)}$.

Proposition 2.4. With the above notations, the non-orthogonal sums $\oplus\limits_k X_k^L$ and $\oplus\limits_k X_k^R$ are dense in $\mathcal{H}_{(n)}$.

Proof. Since $\bigcup V_k$ is dense in $\mathcal{H}_{(n)}$ it suffices to show that each V_k is included in $Y_k^L :=$ the closure of $\oplus\limits_{l \leq k} X_l^L$ in $\mathcal{H}_{(n)}$. Let v be an arbitrary element of V. For an arbitrary, non-negative integer N we successively decompose V_k as

$$V_k = X_{k-1}^L \oplus V_{k-1} = X_{k-1}^L \oplus X_{k-2}^L \oplus V_{k-2} = ... = X_{k-1}^L \oplus ... \oplus X_{k-N}^L \oplus V_{k-N},$$

and write $v = x_1 + x_2 + ... + x_N + v_N$ with $x_j \in X_{k-j}^L$ for $1 \leq j \leq N$ and $v_N \in V_{k-N}$. If we can prove that $v_N \longrightarrow 0$ weakly in $\mathcal{H}_{(n)}$ as N tends to $+\infty$, then we get

$$v = \sum_{j=1}^{\infty} x_j \in \text{ the weak closure of } Y_k^L \text{ in } \mathcal{H}_{(n)},$$

i.e. $v \in Y_k^L$ as desired, by the Hahn-Banach theorem. But

$$\delta \|v_N\|_{(n)} \leq \sup_{\substack{w \in V_{k-N} \\ \|w\|_{(n)} \leq 1}} |B(v_N, w)| = \sup_{\substack{w \in V_{k-N} \\ \|w\|_{(n)} \leq 1}} |B(v, w)| \lesssim \|v\|_{(n)},$$

since $B(X_{k-j}^L, V_{k-N}) = 0$ for $j \leq N$, so that $\|v_N\|_{(n)} \leq \text{const} < +\infty$ for all N. Now the limit ξ of any weakly convergent subsequence of $\{v_N\}_N$ has the property that $\xi \in V_{k-N}$ for all N, therefore $\xi \in \bigcap\limits_N V_{k-N} = \{0\}$, and we are done. ∎

Definition 2.5. Let V a closed left-(or right-)submodule of $\mathcal{H}_{(n)}$. Call a system of vectors $\{v_j\}_j$ in V a left-(or right-)Riesz basis for V if the mapping

$$\{\alpha_j\}_j \longmapsto \sum_j \alpha_j v_j \quad \left(\text{or } \{\alpha_j\}_j \longmapsto \sum_j v_j \alpha_j, \text{ respectively} \right) \tag{2.1}$$

is a continuous isomorphism between $\ell_{\mathbb{N}}^2 \otimes \mathbb{C}_{(n)}$ and V.

Note that the quality of being a left-(or right-)Riesz basis is preserved under the action of a continuous isomorphism of Clifford modules.

Given a CMRA of $\mathcal{H}_{(n)}$, our goal is to construct (if possible) a pair of systems of vectors

$$\{\Theta_{j,k}^L\}_{j,k} \text{ and } \{\Theta_{j,k}^R\}_{j,k},$$

called *dual pair of wavelet bases*, having the next properties:

(1) For each k, $\Theta_{j,k}^L$ belongs to X_k^L for any j;

(2) For each k, $\Theta_{j,k}^R$ belongs to X_k^R for any j;

(3) $\{\Theta_{j,k}^L\}_{j,k}$ is a left-Riesz basis for $\mathcal{H}_{(n)}$;

(4) $\{\Theta_{j,k}^R\}_{j,k}$ is a right-Riesz basis for $\mathcal{H}_{(n)}$;

(5) They are dual to each other with respect to B, i.e.

$$B(\Theta_{j,k}^L, \Theta_{j',k'}^R) = \delta_{j,j'}\delta_{k,k'}, \quad \text{for all } j, k, j', k'.$$

In particular, these conditions will imply that any x in $\mathcal{H}_{(n)}$ has the representations

$$x = \sum_{j,k} B(x, \Theta_{j,k}^R)\Theta_{j,k}^L = \sum_{j,k} \Theta_{j,k}^R B(\Theta_{j,k}^L, x),$$

and, moreover, that

$$\|x\|_{(n)}^2 \approx \sum_{j,k} |B(x, \Theta_{j,k}^R)|^2 \approx \sum_{j,k} |B(\Theta_{j,k}^L, x)|^2.$$

With the same notation as in the definition of CMRA, we define the following (possible unbounded) operators

$$T^L, \ T^R : \bigoplus_k W_k \longrightarrow \mathcal{H}_{(n)},$$

setting

$$T^L := \bigoplus_k \pi_k^L, \quad \text{and} \quad T^R := \bigoplus_k \pi_k^R.$$

Their importance is emphasized by the next result.

Proposition 2.6. *Suppose that for a given CMRA there exist a left-Riesz basis* $\{\Theta_{j,k}^L\}_j$ *in* X_k^L *and a right-Riesz basis* $\{\Theta_{j,k}^R\}_j$ *in* X_k^R, *both uniformly in* k *(in the sense that the norms of the isomorphisms in (2.1) and the norms of their inverses are uniformly bounded in* k), *for which*

$$B(\Theta_{j,k}^L, \Theta_{j',k'}^R) = \delta_{j,j'}\delta_{k,k'}, \quad \text{for all } j, k, j', k'.$$

Then $\{\Theta_{j,k}^L\}_{j,k}$ *and* $\{\Theta_{j,k}^R\}_{j,k}$ *are a dual pair of wavelet bases for this CMRA if and only if* T^L *and* T^R *are bounded operators on* $\mathcal{H}_{(n)}$.

Proof. Taking $x \in \mathcal{H}_{(n)}$,

$$x = \sum_{j,k} \alpha_{j,k}\Theta_{j,k}^L = \sum_k \left(\sum_j \alpha_{j,k}\Theta_{j,k}^L \right) = \sum_k x_k,$$

with $x_k := \sum_j \alpha_{j,k}\Theta_{j,k}^L \in X_k^L$, we have $\|x_k\|_{(n)}^2 \approx \sum_j |\alpha_{j,k}|^2$, uniformly in k. Next, there exists $y = \bigoplus_k y_k \in \bigoplus_k W_k$ such that $\pi_k^L y_k = x_k$ for all k, hence $\|x_k\|_{(n)} \approx \|y_k\|_{(n)}$ uniformly in k, and

$$\|y\|_{(n)}^2 = \sum_k \|y_k\|_{(n)}^2 \approx \sum_k \|x_k\|_{(n)}^2 \approx \sum_k \sum_j |\alpha_{j,k}|^2.$$

Moreover, $T^L y = x$ so that T^L is continuous if and only if

$$\left\| \sum_{j,k} \alpha_{j,k}\Theta_{j,k}^L \right\|_{(n)}^2 \lesssim \sum_{j,k} |\alpha_{j,k}|^2. \tag{2.2}$$

Analogously, T^R is continuous if and only if

$$\left\| \sum_{j,k} \Theta_{j,k}^R \alpha_{j,k} \right\|_{(n)}^2 \lesssim \sum_{j,k} |\alpha_{j,k}|^2. \tag{2.3}$$

The fact that (2.2) and (2.3) also imply the reverse inequalities, and therefore $\{\Theta_{j,k}^L\}_{j,k}$ and $\{\Theta_{j,k}^R\}_{j,k}$ are shown to be left- and right-, respectively, Riesz bases for $\mathcal{H}_{(n)}$, is obtained from a standard duality argument. Writing

$$\left| \sum_{j,k} \alpha_{j,k} \overline{\beta_{j,k}} \right| = \left| B \left(\sum_{j,k} \alpha_{j,k} \Theta_{j,k}^L, \sum_{j,k} \Theta_{j,k}^R \overline{\beta_{j,k}} \right) \right|$$

$$\lesssim \left\| \sum_{j,k} \alpha_{j,k} \Theta_{j,k}^L \right\|_{(n)} \left\| \sum_{j,k} \Theta_{j,k}^R \overline{\beta_{j,k}} \right\|_{(n)}$$

$$\lesssim \left(\sum_{j,k} |\beta_{j,k}|^2 \right)^{1/2} \left\| \sum_{j,k} \alpha_{j,k} \Theta_{j,k}^L \right\|_{(n)}.$$

and taking the supremum of both sides when $\sum |\beta_{j,k}|^2 = 1$, the conclusion follows. ∎

§2.3 BASES IN THE WAVELET SPACES

Consider $\{V_k\}_k$ a CMRA of $\mathcal{H}_{(n)}$ such that V_k is self-adjoint for each $k \in \mathbb{Z}$ (and, hence, so are the corresponding $\{W_k\}_k$). Also, assume that we are given a family of vectors $\{\psi_{j,k}\}_{j,k}$ for which:

(1) $\{\psi_{j,k}\}_j$ live in W_k for each k;

(2) $\overline{\psi_{j,k}} = \psi_{j,k}$ for all j, k;

(3) For any k, $\langle \psi_{j,k}, \psi_{j',k} \rangle = \delta_{j,j'}$ for all j, j';

(4) For any $x \in W_k$, $x = \sum_j \langle x, \psi_{j,k} \rangle \psi_{j,k} = \sum_j \psi_{j,k} \langle \psi_{j,k}, x \rangle$;

(5) For any $x \in W_k$, $\|x\|_{(n)}^2 \approx \sum_j |\langle x, \psi_{j,k} \rangle|^2$.

Note that actually there always exists a family of vectors with the above properties.

Set $\theta_{j,k}^L := \pi_k^L \psi_{j,k} \in X_k^L$ and $\theta_{j,k}^R := \pi_k^R \psi_{j,k} \in X_k^R$, for all j, k. Thus, for each k, $\{\theta_{j,k}^L\}_j$ is a left-Riesz basis in X_k^L, while $\{\theta_{j,k}^R\}_j$ is a right-Riesz basis in X_k^R. Consequently, it makes sense to define a family of bounded operators

$$C_k : X_k^L \longrightarrow X_k^R, \quad k \in \mathbb{Z},$$

by setting

$$C_k \left(\sum_j \alpha_j \theta_{j,k}^L \right) := \sum_j \theta_{j,k}^R \overline{\alpha_j}.$$

Finally, we introduce a $\mathbb{C}_{(n)}$-sesquilinear form on X_k^L, denoted $\{\cdot, \cdot\}$, by putting

$$\left\{ \sum_j \alpha_j \theta_{j,k}^L, \sum_j \beta_j \theta_{j,k}^L \right\} := \sum_j \alpha_j \overline{\beta_j}. \tag{2.4}$$

23

Again from the fact that $\{\theta_{j,k}^L\}_j$ is a left-Riesz basis for X_k^L, we infer that this form is well-defined, continuous, symmetric and non-degenerate on X_k^L. In addition, Re $\{\cdot,\cdot\}$ becomes an inner product on X_k^L which yields the same topology as the one inherited from $\mathcal{H}_{(n)}$. Next, for each $k \in \mathbb{Z}$, consider the $\mathbb{C}_{(n)}$-sesquilinear form $B(\cdot, C_k(\cdot))$ on X_k^L, and note that for any $y \in X_k^L$

$$\sup_{\substack{x \in X_k^L \\ \|x\|_{(n)} \leq 1}} |B(x, C_k y)| \approx \sup_{\substack{x \in V_{k+1} \\ \|x\|_{(n)} \leq 1}} |B(x, C_k y)| \gtrsim \delta \|C_k y\|_{(n)} \approx \delta \|y\|_{(n)}.$$

Here we have used $B(X_k^L, V_k) = 0$ plus the δ-non-degeneracy of B on V_{k+1}. With a similar trick

$$\sup_{\substack{x \in X_k^L \\ \|x\|_{(n)} \leq 1}} |B(y, C_k x)| \approx \sup_{\substack{x \in X_k^R \\ \|x\|_{(n)} \leq 1}} |B(y, x)| \gtrsim \delta \|y\|_{(n)},$$

i.e. this is a δ-non-degenerate form on X_k^L. Thus, by Corollary 1.12 there exists a continuous automorphism U_k of X_k^L such that

$$\{U_k x, y\} = B(x, C_k y), \quad \text{for all } x, y \in X_k^L. \tag{2.5}$$

Lemma 2.7. *If the form B is δ-accretive on $\mathcal{H}_{(n)}$, then U_k is δ-accretive on X_k^L endowed with the inner product Re $\{\cdot,\cdot\}$.*

Proof. The key observation to be made here is that $C_k x - \overline{x} \in V_k$ for any $x \in X_k^L$. This is because if $x = \sum_j \alpha_j \theta_{j,k}^L$, then

$$C_k x - \overline{x} = \sum_j \theta_{j,k}^R \overline{\alpha_j} - \sum_j \overline{\theta_{j,k}^L \alpha_j} = \sum_j (\theta_{j,k}^R - \psi_{j,k}) \overline{\alpha_j} + \overline{\sum_j \alpha_j (\psi_{j,k} - \theta_{j,k}^L)} \in V_k,$$

by the self-adjointness of $\psi_{j,k}$'s and V_k. Thus, for any $x \in X_k^L$, using $B(X_k^L, V_k) = 0$ we get

$$\text{Re}\, \{U_k x, x\} = \text{Re}\, B(x, C_k x) = \text{Re}\, B(x, \overline{x}) \geq \delta \|x\|_{(n)}^2,$$

where the last inequality follows from the δ-accretivity of B. ∎

Proposition 2.8. *If for each k we set $\Theta_{j,k}^L := U_k^{-1} \theta_{j,k}^L$ and $\Theta_{j,k}^R := \theta_{j,k}^R$, then $\{\Theta_{j,k}^L\}_j$ is a left-Riesz basis for X_k^L, $\{\Theta_{j,k}^R\}_j$ is a right-Riesz basis for X_k^R, both uniformly in k, so that*

$$B(\Theta_{j,k}^L, \Theta_{j',k'}^R) = \delta_{j,j'} \delta_{k,k'}, \quad \text{for all } j, k, j', k'.$$

24

Proof. The first part of the statement follows from the fact that for any $k \in \mathbb{Z}$

$$0 < c^{-1} \leq \|\pi_k^L\|, \ \|\pi_k^R\|, \ \|U_k^{-1}\| \leq c < +\infty,$$

for some constants c depending solely on n, $\|B\|$ and δ. Further, by (2.4),

$$B(\Theta_{j,k}^L, \Theta_{j',k}^R) = B(U_k^{-1}\theta_{j,k}^L, C_k\theta_{j',k}^L) = \{\theta_{j,k}^L, \theta_{j',k}^L\} = \delta_{j,j'}.$$

Finally, the B-orthogonality between various levels in k is easily seen from $X_k^L \subset V_{k+1}$ and $B(V_{k+1}, X_{k'}^R) = 0$ if e.g. $k' \geq k+1$, etc. $\blacksquare$

We conclude this section with the following technical result which is needed later.

Proposition 2.9. *Let V be a closed left-submodule of $\mathcal{H}_{(n)}$, B a $\mathbb{C}_{(n)}$-sesquilinear form on V, and $\{v_j\}_j$ a left-Riesz basis of V so that*

$$\mathcal{B}(v_j, v_{j'}) = \delta_{j,j'}, \quad \text{for all } j, j'. \tag{2.6}$$

If V is equipped with the inner product $\operatorname{Re} \mathcal{B}(\cdot, \cdot)$ and T is a δ-accretive automorphism of $(V, \operatorname{Re} \mathcal{B})$ such that $|\mathcal{B}(Tv_j, v_k)| \lesssim \exp(-\alpha|j-k|)$ for some $\alpha > 0$ and all j, k, then, for some $\alpha' > 0$, we also have

$$|\mathcal{B}(T^{-1}v_j, v_k)| \lesssim \exp(-\alpha'|j-k|), \quad \text{for all } j, k.$$

Proof. Let us first remark that (2.6) implies that $\mathcal{B}$ is non-degenerate and symmetric, therefore $\operatorname{Re}\mathcal{B}(\cdot, \cdot)$ is, indeed, an inner product on V. Moreover, for $x = \sum_j \alpha_j v_j \in V$, one has $\operatorname{Re}\mathcal{B}(x, x) = \sum_j \operatorname{Re}\alpha_j\overline{\alpha_j} = \sum_j |\alpha_j|^2 \approx \|x\|_{(n)}^2$.

Now, for a fixed, large $M > 0$, we write

$$T^{-1} = \frac{1}{M}\left(I + \frac{T - M}{M}\right)^{-1} = \frac{1}{M}\sum_{k=0}^{\infty}(-1)^k\left(\frac{T - M}{M}\right)^k.$$

The series is absolutely convergent in the strong operator norm on $(V, \operatorname{Re}\mathcal{B}(\cdot, \cdot))$ since, for suitably large M, $\omega := \|(T - M)/M\| < 1$, as one can readily check using the δ-accretivity of T. Let now $\{t_k(j, j')\}_{j,j'}$ be the matrix representation of

$(T-M)^{k+1}M^{-(k+1)}$, $k \geq 0$, with respect to the orthonormal basis $\{v_j\}_j$ in $(V, \operatorname{Re} \mathcal{B})$. Inductively, we see that

$$t_k(j,j') = \sum_{j_1} \cdots \sum_{j_{k-1}} t_0(j,j_1)\, t_0(j_1,j_2)\ldots t_0(j_{k-1},j'). \qquad (2.7)$$

Using this together with $|t_0(j,j')| \lesssim \exp(-\alpha|j - j'|)$, we conclude that there exist some positive constants C and β such that

$$|t_k(j,j')| \leq C^k \exp(-\beta|j - j'|), \quad \text{for all } k,j,j'. \qquad (2.8)$$

On the other hand,

$$|t_k(j,j')| = \left| \operatorname{Re} \mathcal{B}\left(\left(\frac{T-M}{M}\right)^k v_j, v_{j'} \right) \right| \leq \omega^k, \qquad (2.9)$$

so that a logarithmically convex combination of (2.8) and (2.9) shows that there exist $\gamma > 0$ and $0 < \lambda < 1$ for which

$$|t_k(j,j')| \lesssim \lambda^k \exp(-\gamma|j - j'|), \quad \text{for all } k,j,j'. \qquad (2.10)$$

Consequently, returning to (2.7), (2.10) gives us

$$|\mathcal{B}(T^{-1}v_j, v_{j'})| \lesssim \sum_k \left| \mathcal{B}\left(\left(\frac{T-M}{M}\right)^k v_j, v_{j'} \right) \right| \approx \sum_k |t_k(j,j')|$$

$$\lesssim \left(\sum_{k=0}^{\infty} \lambda^k \right) \exp(-\gamma|j - j'|) \approx \exp(-\gamma|j - j'|),$$

and the proof is complete. ∎

§2.4 Clifford Multiresolution Analyses of $L^2(\mathbb{R}^m) \otimes \mathbb{C}_{(n)}$

The particular context we shall discuss in this section is $\mathcal{H} := L^2(\mathbb{R}^m)$, with the involution given by the usual conjugation of complex-valued functions, and

$$B(f,g) := \int_{\mathbb{R}^m} f(x)b(x)g(x)\, dx, \quad f,g \in L^2(\mathbb{R}^m)_{(n)}, \qquad (2.11)$$

where $b : \mathbb{R}^m \longrightarrow \mathbb{R}^{n+1} \subset \mathbb{C}_{(n)}$ is a L^∞-function with $\mathrm{Re}\, b(x) \geq \delta > 0$. Note that, according to Proposition 2.1, B is a δ-accretive form on $L^2(\mathbb{R}^m)_{(n)}$.

Consider now $\{V'_k\}_k$ a *multiresolution analysis* of $L^2(\mathbb{R}^m)$ ([Me]), that is, a family $\{V'_k\}_k$ of closed subspaces of $L^2(\mathbb{R}^m)$ for which:

(1) $\cap_{-\infty}^{+\infty} V'_k = \{0\}$ and $\cup_{-\infty}^{+\infty} V'_k$ is dense in $L^2(\mathbb{R}^m)$;

(2) For any $k \in \mathbb{Z}$, $f(x) \in V'_k \Longleftrightarrow f(2x) \in V'_{k+1}$;

(3) For any $j \in \mathbb{Z}$, $f(x) \in V'_k \Longrightarrow f(x - j) \in V'_k$;

(4) There exists $\phi(x) \in V'_0$ such that $\{\phi(x - j)\}_j$ is an orthonormal basis for V'_0.

We make the supplementary assumptions that $\phi \in C^r(\mathbb{R}^m)$ for some nonnegative integer r, and that all its partial derivatives have exponential decay at infinity, i.e. there exists a certain constant $\varkappa > 0$ so that

$$|\partial^\alpha \phi(x)| \lesssim \exp\left(-\varkappa |x|\right), \quad x \in \mathbb{R}^m,$$

for any multi-index α having $|\alpha| \leq r$.

From the standard theory (see [Me]), let us recall that the functions $\{\phi_{j,k}\}_j$ given by

$$\phi_{j,k}(x) := 2^{km/2} \phi(2^k x - j), \quad k \in \mathbb{Z}, \ j \in \mathbb{Z}^m, \qquad (2.12)$$

form an orthonormal basis for V'_k, and that there exist $2^m - 1$ functions $\{\psi_\epsilon\}_\epsilon$ in V'_1, having the same type of regularity and decay as ϕ, which form an orthonormal basis for the wavelet space $W'_0 := V'_1 \ominus V'_0$. In particular, $\{\psi_{\epsilon,j,k}\}_{\epsilon,j}$ with

$$\psi_{\epsilon,j,k}(x) := 2^{km/2} \psi_\epsilon(2^k x - j), \quad k \in \mathbb{Z}, \ j \in \mathbb{Z}^m, \qquad (2.13)$$

is an orthonormal basis for $W'_k := V'_{k+1} \ominus V'_k$. In the sequel, we shall identify ϕ with $\phi \otimes e_0$, ψ_ϵ with $\psi_\epsilon \otimes e_0$, etc. We shall also assume that $\overline{V'_k} = V'_k$ for any $k \in \mathbb{Z}$.

Setting $V_k := V'_k \otimes \mathbb{C}_{(n)}$, $W_k := W'_k \otimes \mathbb{C}_{(n)}$, and then taking $\{V_k\}_k$ together with the form (2.11), we then obtain a self-adjoint CMRA for $L^2(\mathbb{R}^m)_{(n)}$. The main result of this section concerns the existence and regularity of a dual pair of wavelet bases for this CMRA.

Theorem 2.10. *For the above CMRA of $L^2(\mathbb{R}^m)_{(n)}$ there exists a dual pair of wavelet bases $\{\Theta^L_{\epsilon,j,k}\}_{\epsilon,j,k}$ and $\{\Theta^R_{\epsilon,j,k}\}_{\epsilon,j,k}$ which are r-regular in the sense that*

$$\Theta^L_{\epsilon,j,k}, \ \Theta^R_{\epsilon,j,k} \in C^r(\mathbb{R}^m)_{(n)}, \quad \text{for all } j, k, \epsilon,$$

27

and there exists some $\kappa > 0$ so that

$$|\partial^\alpha \Theta^L_{\epsilon,j,k}(x)| + |\partial^\alpha \Theta^R_{\epsilon,j,k}(x)| \lesssim 2^{k(m/2+|\alpha|)} \exp(-\kappa|2^k x - j|),$$

for all j, k, ϵ and all multi-indices α with $|\alpha| \leq r$.

Proof. Starting with the functions $\{\psi_{\epsilon,j,k}\}_{\epsilon,j,k}$ from (2.13), the algorithm presented in the previous section allows us to construct two families of (left- and right-, respectively) Riesz bases, $\{\Theta^L_{\epsilon,j,k}\}_{\epsilon,j}$ in X^L_k and $\{\Theta^R_{\epsilon,j,k}\}_{\epsilon,j}$ in X^R_k, both uniformly in k, for which

$$B(\Theta^L_{\epsilon,j,k}, \Theta^R_{\epsilon',j',k'}) = \delta_{\epsilon,\epsilon'} \delta_{j,j'} \delta_{k,k'}.$$

Now we use a version of the aforementioned algorithm, this time starting with $\{\phi_{j,k}\}_{j,k}$ (from (2.12)), to produce for each fixed $k \in \mathbb{Z}$ a left-Riesz basis $\{\phi^L_{j,k}\}_j$ and a right-Riesz basis $\{\phi^R_{j,k}\}_j$ for V_k such that

$$B(\phi^L_{j,k}, \phi^R_{j',k}) = \delta_{j,j'}.$$

More specifically, we can take $\phi^R_{j,k} := \phi_{j,k}$ and $\phi^L_{j,k} := S^{-1}_k \phi_{j,k}$ where S_k is the unique continuous left-Clifford-linear operator $S_k : V_k \longrightarrow V_k$ such that $\langle S_k f, g \rangle = B(f, \overline{g})$, for all f, g in V_k (S_k is the analog of U_k from (2.5)).

Since

$$\phi^L_{j,k} = \sum_l \langle S^{-1}_k \phi_{j,k}, \phi_{l,k} \rangle \phi_{l,k},$$

a simple application of Proposition 2.9 shows that $\{\phi^L_{j,k}\}_{j,k}$ and $\{\phi^R_{j,k}\}_{j,k}$ have the same smoothness and decay as the initial ϕ. Moreover,

$$\pi^L_k f = f - \sum_j B(f, \phi^R_{j,k}) \phi^L_{j,k}, \quad \text{and} \quad \pi^R_k f = f - \sum_j \phi^R_{j,k} B(\phi^L_{j,k}, f), \quad f \in W_k.$$

Returning now to our old Θ's, recall that in fact we can take

$$\Theta^R_{\epsilon,j,k} := \theta^R_{\epsilon,j,k} = \pi^R_k \psi_{\epsilon,j,k} = \psi_{\epsilon,j,k} - \sum_l \phi^R_{l,k} B(\phi^L_{l,k}, \psi_{\epsilon,j,k}),$$

so that the regularity and decay properties of $\Theta^R_{\epsilon,j,k}$ immediately follow from the corresponding ones for $\psi_{\epsilon,j,k}$'s, $\phi^L_{j,k}$'s and $\phi^R_{j,k}$'s. As for $\Theta^L_{\epsilon,j,k} := U^{-1}_k \theta^L_{\epsilon,j,k} =$

$U_k^{-1}(\pi_k^L \psi_{\epsilon,j,k})$ (recall that U_k has been introduced in (2.5)), a similar argument holds, although we have to invoke Proposition 2.9 one more time (all technicalities have been taken care of in the previous section). According to Proposition 2.6, all that remains to be proved is the boundedness of the operators T^L, T^R. Note that the distribution kernel of e.g. T^L is

$$K(x,y) := \sum_{j,k} \sum_\epsilon \theta_{\epsilon,j,k}^L(x)\psi_{\epsilon,j,k}^L(y), \quad x \neq y.$$

This is easily checked to be a standard kernel, therefore the L^2-boundedness of T^L can be obtained using a Clifford algebra version of the celebrated T(1) theorem of David and Journé [DJ] (see also the next chapter). However, the computations are completely analogous to those for the scalar case (see [Me] and [Tc]), hence we omit them. ∎

Corollary 2.11. *With the above notations, for all j, k, ϵ we have*

$$\int_{\mathbb{R}^m} \Theta_{\epsilon,j,k}^L(x)b(x)\,dx = \int_{\mathbb{R}^m} b(x)\Theta_{\epsilon,j,k}^R(x)\,dx = 0.$$

Proof. The constant function 1 belongs to the $L^2(\mathbb{R}^m, e^{-\kappa|x|}dx)_{(n)}$-closure of V_k and, consequently, everything follows from $B(X_k^L, V_k) = B(V_k, X_k^R) = 0$. ∎

Exercise. Let $Q_{j,k}$ stand for the dyadic cube $\{x \in \mathbb{R}^n \,;\, 2^k x - j \in [0,1]^n\}$, and let $H^1(\mathbb{R}^n)$ stand for the usual Hardy space (see e.g. [St]).

Prove that for a sequence $\{c_{\epsilon,j,k}\}_{\epsilon,j,k}$ of elements from $\mathbb{C}_{(n)}$ the following are equivalent:

(1) $A := \left(\sum_{j\in\mathbb{Z}^n} \sum_{k\in\mathbb{Z}} \sum_\epsilon 2^{nk}|c_{cj,k}|^2 \chi_{Q_{j,k}}\right)^{1/2} \in L^1(\mathbb{R}^n)$;

(2) $B := \sum_{j\in\mathbb{Z}^n} \sum_{k\in\mathbb{Z}} \sum_\epsilon c_{\epsilon,j,k}\Theta_{\epsilon,j,k}^L \in bH^1(\mathbb{R}^n)_{(n)}$;

(3) $C := \sum_{j\in\mathbb{Z}^n} \sum_{k\in\mathbb{Z}} \sum_\epsilon \Theta_{\epsilon,j,k}^R c_{\epsilon,j,k} \in bH^1(\mathbb{R}^n)_{(n)}$;

Moreover, if the above conditions are fulfilled, then $\|A\|_{L^1} \approx \|B\|_{bH^1} \approx \|C\|_{bH^1}$. In particular, $\{\Theta_{\epsilon,j,k}^L\}_{\epsilon,j,k}$ and $\{\Theta_{\epsilon,j,k}^R\}_{\epsilon,j,k}$ are unconditional basis for $bH^1(\mathbb{R}^n)_{(n)}$.

Remarks.

(1) Since $X_k^L, X_k^R \subseteq V_{k+1}$, we have that $\Theta_{\epsilon,j,k}^L, \Theta_{\epsilon,j,k}^R \in V_{k+1}' \otimes \mathbb{C}_{(n)}$ for all ϵ, j, k.

(2) Using the exponential decay of the Θ's as before, we can get higher order vanishing moments for Θ's provided the initial multiresolution analysis is

suitably chosen. If we take V_0' to be e.g. the m-fold tensor product of the compactly supported real spline functions of order $r+2$ in $L^2(\mathbb{R})$ having integer breakpoints, then we have

$$\int_{\mathbb{R}^m} x^\alpha \Theta_{\epsilon,j,k}^L(x)\, b(x) dx = 0 \text{ and } \int_{\mathbb{R}^m} x^\alpha b(x) \Theta_{\epsilon,j,k}^R(x)\, dx = 0, \quad \forall\, |\alpha| \le r+2.$$

(3) The same results continue to hold if the exponential decay is replaced with a rapid decay.

Finally, let us mention that the main theorem of this section can be adapted to contain the case of a dyadic pseudo-accretive function b, i.e. a L^∞, $\mathbb{R}^{n+1}$−valued function whose integral means over dyadic cubes are greater than a certain fixed, positive δ. More specifically, we note the following result from [AT].

Theorem 2.12. *For any dyadic pseudo-accretive function b in $\mathbb{R}^m$, there exists a CMRA of $L^2(\mathbb{R}^m)_{(n)}$ with $B(\cdot,\cdot)$ given by (2.11) for which one can construct a dual pair of wavelet bases $\{\Theta_{\epsilon,j,k}^L\}$, $\{\Theta_{\epsilon,j,k}^R\}$ with small regularity, i.e. for some $0 < r < 1$, one has that for any $N \in \mathbb{N}$, there exists $C_N > 0$ such that, for all j, k, ϵ,*

$$|\Theta_{\epsilon,j,k}^L(x)| \le C_N 2^{km/2}(1 + |2^k x - j|)^{-m-N}, \quad x \in \mathbb{R}^m,$$

$$|\Theta_{\epsilon,j,k}^L(x) - \Theta_{\epsilon,j,k}^L(y)|$$
$$\le C_N 2^{k(m/2+r)}|x-y|^r((1 + |2^k x - j|)^{-m-N} + (1 + |2^k y - j|)^{-m-N}),$$

$x, y \in \mathbb{R}^m$, *and similar estimates for $\Theta_{\epsilon,j,k}^R$'s.*

The idea is to start with a very simple CMRA, e.g. the one associated to the Haar system. One cannot conclude here because of the lack of regularity of the Haar system, but a suitable perturbation of this special case will do.

§2.5 HAAR CLIFFORD WAVELETS

First let us introduce some notation. For any $k \in \mathbb{Z}$ let $\mathcal{F}_k$ denote the collection of all dyadic cubes Q,

$$Q = Q_{k,v} = \{x \in \mathbb{R}^m\,;\, 2^{-k}v_i \le x_i \le 2^{-k}(v_i + 1),\quad i = 1, 2, ..., m\}, \quad v \in \mathbb{Z}^m,$$

with side-length $l(Q) := 2^{-k}$, and set $\mathcal{F} := \bigcup_{k \in \mathbb{Z}} \mathcal{F}_k$. Each dyadic cube $Q \in \mathcal{F}$ has 2^m "children"

$$\{Q^j\}_{j=1}^{2^m} := \{Q' \in \mathcal{F}_{k+1} \,;\, Q' \subset Q\}.$$

For Q cube and λ positive constant, λQ will stand for the cube having the same center as Q and side-length $\lambda l(Q)$. We also set χ_Q for the characteristic function of Q.

The CMRA of $L^2(\mathbb{R}^m)_{(n)}$ we shall work with throughout this section consists of

$$V_k := \{f \in L^2(\mathbb{R}^m)_{(n)} \,;\, f \text{ piecewise constant on the dyadic cubes of } \mathcal{F}_k\},$$

for $k \in \mathbb{Z}$, together with the δ-accretive form $B(\cdot, \cdot)$ given by (2.11). This time, following [CJS], we shall produce explicit expressions for a system of Clifford wavelets having 0-regularity, i.e. a system of *Haar Clifford wavelets*. In fact, for this particular context, we can perform our construction in a slightly more general form. Suppose that, for some $\delta > 0$, the L^∞–function $b : \mathbb{R}^m \longrightarrow \mathbb{R}^{n+1}$ from the definition of $B(\cdot, \cdot)$ is actually δ−*dyadic pseudo-accretive*, i.e. it satisfies

$$\left| \frac{1}{|Q|} \int_Q b(x)\, dx \right| \geq \delta, \tag{2.14}$$

for any dyadic cube Q in $\mathbb{R}^m$; here $|Q|$ denotes the Euclidean volume of Q. Note that in this case, by Lebesgue's differentiation theorem, one has $\|b^{-1}\|_{L^\infty} \leq \delta^{-1}$.

Next, we introduce

$$m(Q) := \int_Q b(x)\, dx, \quad Q \in \mathcal{F}.$$

Our hypotheses on b imply that $m(Q) \in \mathbb{R}^{n+1}$ and $|m(Q)| \approx |Q|$. For each $Q \in \mathcal{F}$ we first construct a family of $2^m - 1$ functions in V_{k+1}, denoted by $\{\theta_{Q,i}\}_{i=1}^{2^m-1}$, such that

(1) $\int_{\mathbb{R}^m} \theta_{Q,i}(x) b(x)\, dx = \int_{\mathbb{R}^m} b(x) \theta_{Q,i}(x)\, dx = 0$, $\quad i = 1, 2, ..., 2^m - 1$;

(2) $\int_{\mathbb{R}^m} \theta_{Q,i}(x) b(x) \theta_{Q,j}(x)\, dx = \delta_{ij}$ for all i, j.

Actually we shall take

$$\theta_{Q,i} := a_i \left(\sum_{j=1}^{i} \chi_{Q^j} \right) - b_{i+1} \chi_{Q^{i+1}}, \tag{2.15}$$

31

for some a_i, $b_i \in \mathbb{C}_{(n)}$, $i = 1, 2, ..., 2^m - 1$, suitably chosen. It is visible from (2.15) that unless $\theta_{Q,i}$ and $\theta_{Q,j}$ have the same pair of subscripts, one of them is constant on the support of the other one. Thus, (1) automatically implies (2), at least for $i \neq j$. However, (1) is fulfilled if we choose

$$a_i := \left(\sum_{j=1}^{i} m(Q^j) \right)^{-1} \quad \text{and} \quad b_{i+1} := m(Q^{i+1})^{-1}, \quad \text{for } i = 1, \ldots, 2^m - 1,$$

if we have $\sum_{j=1}^{i} m(Q^j) \neq 0$, for $i = 1, 2, ..., 2^m - 1$. This is taken care of in the following elementary lemma.

Lemma 2.13. *Consider N vectors in a normed vector space $(V, \|\cdot\|)$ and let S denote the norm of their sum. Then there exists an enumeration of them, say $v_1, v_2, ..., v_N$, so that $\|v_1 + v_2 + ... + v_i\| \geq S/N$ for $i = 1, 2, ..., N$.*

Proof. We proceed inductively. Let $w_1, w_2, ..., w_N$ be an arbitrary enumeration of the given family of vectors. Since

$$\sum_{i=1}^{N} \left\| \sum_{j \neq i} w_j \right\| \geq \left\| \sum_{i=1}^{N} (\sum_{j \neq i} w_j)' \right\| = (N-1) \left\| \sum_{k=1}^{N} w_k \right\| = (N-1)S,$$

we infer the existence of an index i_0 for which

$$S' := \left\| \sum_{j \neq i_0} w_j \right\| \geq \frac{N-1}{N} S.$$

For $\{w_j\}_{j \neq i_0}$ we use the induction hypothesis and get an enumeration $\{w_j\}_{j \neq i_0} = \{v_i\}_{i=1}^{N-1}$ such that $\|v_1 + v_2 + ... + v_i\| \geq S'/(N-1) \geq S/N$ for $i = 1, 2, ..., N-1$. All we have to do now is to rename w_{i_0} to be v_N. ∎

Since in our situation

$$\left| \sum_{j=1}^{2^m} m(Q^j) \right| = |m(Q)| \approx |Q|,$$

it follows that one can enumerate the children of Q such that

$$\left| \sum_{j=1}^{i} m(Q^j) \right| \approx |Q|, \quad \text{for } i = 1, 2, ..., 2^m - 1.$$

32

As for the case $i = j$ in (2), introducing

$$M(Q, i) := \int_{\mathbb{R}^m} \theta_{Q,i}(x) b(x) \theta_{Q,i}(x) \, dx,$$

a direct calculation shows that

$$M(Q, i) = \left(\sum_{j=1}^{i} m(Q^j) \right)^{-1} \left(\sum_{j=1}^{i+1} m(Q^j) \right) m(Q^{i+1})^{-1}.$$

In particular $|M(Q, i)| \approx |Q|^{-1}$, and $M(Q, i)$ is a Clifford vector.

Finally, we the define Haar Clifford wavelets by normalizing the θ's

$$\Theta^L_{Q,i} := M(Q, i)^{-1/2} \theta_{Q,i}, \quad i = 1, 2, ..., 2^m - 1 \qquad (2.16)$$

$$\Theta^R_{Q,i} := \theta_{Q,i} M(Q, i)^{-1/2}, \quad i = 1, 2, ..., 2^m - 1. \qquad (2.17)$$

Note that unless they vanish, $\Theta^L_{Q,i}$ and $\Theta^R_{Q,i}$ take on values in the Clifford group of $\mathbb{R}_{(n)}$.

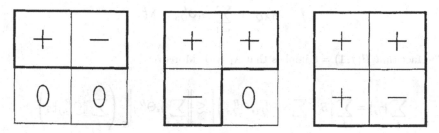

FIGURE 2.1.   The three Haar Clifford wavelets
living in the same dyadic cube for $m = 2$.

The main result of this section is the following.

Theorem 2.14. *With the above hypotheses, $\{\Theta^L_{Q,i}\}_{Q,i}$ and $\{\Theta^R_{Q,i}\}_{Q,i}$ given by (2.16) and (2.17) satisfy:*

(1) $\Theta^L_{Q,i}, \Theta^R_{Q,i} \in V_k$ *for all* $k \in \mathbb{Z}$, $Q \in \mathcal{F}_k$ *and* $i = 1, 2, ..., 2^m - 1$;

(2) $supp\, \Theta^L_{Q,i}, supp\, \Theta^R_{Q,i} \subseteq Q$ *and* $|\Theta^L_{Q,i}|, |\Theta^R_{Q,i}| \lesssim |Q|^{-1/2}$;

(3) $\int_{\mathbb{R}^m} \Theta^L_{Q,i}(x) b(x) \, dx = \int_{\mathbb{R}^m} b(x) \Theta^R_{Q,i}(x) \, dx = 0$ *for all* $Q \in \mathcal{F}$, $i = 1, ..., 2^m - 1$;

(4) $\int_{\mathbb{R}^m} \Theta^L_{Q,i}(x) b(x) \Theta^R_{Q',i'}(x) \, dx = \delta_{Q,Q'} \delta_{i,i'}$, *for all* Q, Q', i, i';

(5) $\{\Theta^L_{Q,i}\}_{Q,i}$ *is a left-Riesz basis for* $L^2(\mathbb{R}^m)_{(n)}$ *and* $\{\Theta^R_{Q,i}\}_{Q,i}$ *is a right-Riesz basis for* $L^2(\mathbb{R}^m)_{(n)}$.

33

Proof. The only thing that we still have to check is (5). For each $k \in \mathbb{Z}$ we consider

$$X_k^L := \{f \in V_{k+1} \,; \int_Q f(x) b(x) \, dx = 0, \text{ for any } Q \in \mathcal{F}_k\},$$

and

$$X_k^R := \{f \in V_{k+1} \,; \int_Q b(x) f(x) \, dx = 0, \text{ for any } Q \in \mathcal{F}_k\}.$$

We claim that $\{\Theta_{Q,i}^L\}$, with $Q \in \mathcal{F}_k$ and $i = 1, 2, ..., 2^m - 1$, is a left-Riesz basis for X_k^L uniformly in $k \in \mathbb{Z}$. Restricting our attention to one dyadic cube $Q \in \mathcal{F}_k$ and using the explicit expressions of the Θ^L's, we readily see that χ_{Q^2} is spanned by χ_{Q^1} and $\Theta_{Q,1}^L$ in the set of $\mathbb{C}_{(n)}$-valued functions on $\mathbb{R}^n$ with its natural structure as a left Clifford module. Continuing this inductively, we see that any characteristic function χ_{Q^i} is spanned by χ_{Q^1} and $\Theta_{Q,1}^L, ..., \Theta_{Q,2^m-1}^L$. Now, if f is the restriction to Q of a function from X_k^L, we have

$$f = \lambda_1 \chi_{Q^1} + \sum_{i=1}^{2^m-1} \gamma_i \Theta_{Q,i}^L \in X_k^L.$$

The fact that $B(f,1) = 0$ implies that $\lambda_1 = 0$. Moreover,

$$\sum_j |\gamma_j| = \sum_j \left| B\left(\sum_i \gamma_i \Theta_{Q,i}^L, \Theta_{Q,j}^R\right)\right| \lesssim \left\|\sum_i \gamma_i \Theta_{Q,i}^L\right\|_2 \left(\sum_j \|\Theta_{Q,j}^R\|_2\right)$$

$$\lesssim \left\|\sum_i \gamma_i \Theta_{Q,i}^L\right\|_2 \lesssim \sum_i |\gamma_i| \|\Theta_{Q,i}^L\|_2 \lesssim \sum_i |\gamma_i|,$$

since $\|\Theta_{Q,i}^L\|_2$, $\|\Theta_{Q,j}^R\|_2 \lesssim 1$. Finally, since there are only finitely many γ_i's, the ℓ^2-sum is comparable with the ℓ^1-sum, so $\|f\|_2^2 \approx \sum_i |\gamma_i|^2$ and this proves the claim. A similar result is valid for Θ^R's also.

At this point, by Proposition 2.6, everything is reduced to proving the estimates:

$$\sum_{Q \in \mathcal{F}} \sum_{i=1}^{2^m-1} |B(\Theta_{Q,i}^L, f)|^2 \lesssim \|f\|_2^2, \quad \text{uniformly for } f \in L^2(\mathbb{R}^m)_{(n)}, \qquad (2.18)$$

$$\sum_{Q \in \mathcal{F}} \sum_{i=1}^{2^m-1} |B(f, \Theta_{Q,i}^R)|^2 \lesssim \|f\|_2^2, \quad \text{uniformly for } f \in L^2(\mathbb{R}^m)_{(n)}. \qquad (2.19)$$

34

To this end, we introduce the projection operators $\triangle_k^L$, $\triangle_k^R$

$$\triangle_k^L : L^2(\mathbb{R}^m)_{(n)} \longrightarrow X_k^L, \quad \triangle_k^R : L^2(\mathbb{R}^m)_{(n)} \longrightarrow X_k^R,$$

by setting

$$\triangle_k^L f := \sum_{Q \in \mathcal{F}_k} \sum_{i=1}^{2^m-1} B(f, \Theta_{Q,i}^R) \Theta_{Q,i}^L, \quad \triangle_k^R f := \sum_{Q \in \mathcal{F}_k} \sum_{i=1}^{2^m-1} \Theta_{Q,i}^R B(\Theta_{Q,i}^L, f). \quad (2.20)$$

Clearly,

$$\|\triangle_k^L f\|_2^2 \approx \sum_{Q \in \mathcal{F}_k} \sum_{i=1}^{2^m-1} |B(f, \Theta_{Q,i}^R)|^2 \quad (2.21)$$

and

$$\|\triangle_k^R f\|_2^2 \approx \sum_{Q \in \mathcal{F}_k} \sum_{i=1}^{2^m-1} |B(\Theta_{Q,i}^L, f)|^2, \quad (2.22)$$

uniformly in $f \in L^2(\mathbb{R}^m)_{(n)}$ and $k \in \mathbb{Z}$.

Next, we consider the so called *conditional expectation operators* (left and right, respectively) E_k^L, E_k^R with respect to the σ-algebra generated by $\mathcal{F}_k$ and the Clifford-algebra valued measure $b(x)dx$, i.e.

$$E_k^L f(x) := \left(\int_Q f(y) b(y)\, dy \right) m(Q)^{-1}, \quad \text{if } x \in Q \in \mathcal{F}_k,$$

and

$$E_k^R f(x) := m(Q)^{-1} \left(\int_Q b(y) f(y)\, dy \right), \quad \text{if } x \in Q \in \mathcal{F}_k,$$

respectively.

The relation between these operators and $\triangle_k^L$, $\triangle_k^R$ is the following.

Lemma 2.15. *We have that* $\triangle_k^L = E_{k+1}^L - E_k^L$ *and* $\triangle_k^R = E_{k+1}^R - E_k^R$.

Proof. By restricting our attention to one dyadic cube $Q \in \mathcal{F}_k$ at a time, we easily see that $B(E_{k+1}^L f - E_k^L f, 1) = 0$ and that $E_{k+1}^L f - E_k^L f$ is constant on each dyadic subcube of Q, i.e. $E_{k+1}^L f - E_k^L f \in X_k^L$. Since $\{\Theta_{Q,i}^L\}$ is a left-Riesz basis for X_k^L, it suffices to show that both $\triangle_k^L f$ and $E_{k+1} f - E_k^L f$ have the same coefficients with respect to this basis, or even that $B(E_{k+1}^l f, \Theta_{Q,j}^R) = B(f, \Theta_{Q,j}^R)$, since $E_k^L f$ is constant on Q.

However, if $\Theta_{Q,j}^R = \sum_i \lambda_i^j \chi_{Q^i}$ and $E_{k+1}^L f = \sum_i f_{Q^i} m(Q^i)^{-1} \chi_{Q^i}$, where we set $f_{Q^i} := |Q^i|^{-1} \int_{Q^i} fb$, then

$$B(E_{k+1}^L f, \Theta_{Q,j}^R) = \sum_i |Q^i| f_{Q^i} \lambda_i^j = \sum_i \int_{Q^i} f b \lambda_i^j \, dx = B(f, \Theta_{Q,j}^R).$$

∎

Before we come to the proof of (2.18) and (2.19), we consider the case $b(x) \equiv 1$ on $\mathbb{R}^m$. For $Q \in \mathcal{F}$ and $i = 1, ..., 2^m - 1$, we set

$$h_Q^i := \frac{2^{m/2}}{|Q|^{1/2}} \left(\frac{i}{i+1}\right)^{1/2} \left\{\frac{1}{i} \sum_{\nu=1}^i \chi_{Q^\nu} - \chi_{Q^{i+1}}\right\}.$$

The family $\{h_Q^i\}_{Q,i}$ is easily checked to be an orthonormal basis for $L^2(\mathbb{R}^m)$ with the standard inner product $\langle \cdot, \cdot \rangle$. In this special case, we denote $\triangle_k^L$ (or $\triangle_k^R$) and E_k^L (or E_k^R) by $\triangle_k$ and E_k, respectively. As in the general case, $\triangle_k = E_{k+1} - E_k$. Moreover, since

$$\triangle_k f = \sum_{Q \in \mathcal{F}_k} \sum_{i=1}^{2^m-1} \langle f, h_Q^i \rangle h_Q^i,$$

we have that

$$\sum_{k=-\infty}^{+\infty} \int_{\mathbb{R}^m} |\triangle_k f|^2 dx = \|f\|_2^2, \qquad f \in L^2(\mathbb{R}^m)_{(n)}. \tag{2.23}$$

Proof of (2.18) and (2.19). Note that $E_k^L f = E_k(fb) E_k(b)^{-1}$. Hence,

$$\begin{aligned}
|\triangle_k^L f| = |E_{k+1}^L f - E_k^L f| &= |E_{k+1}(fb) E_{k+1}(b)^{-1} - E_k(fb) E_k(b)^{-1}| \\
&\leq |E_{k+1}(fb) E_{k+1}(b)^{-1} - E_k(fb) E_{k+1}(b)^{-1}| \\
&\quad + |E_k(fb) E_{k+1}(b)^{-1} - E_k(fb) E_k(b)^{-1}| \\
&\lesssim |\triangle_k(fb)| + |E_k(fb)||E_{k+1}(b)^{-1} - E_k(b)^{-1}| \\
&\lesssim |\triangle_k(fb)| + |E_k(fb)||\triangle_k(b)|,
\end{aligned} \tag{2.24}$$

since $|E_k(b)| \approx 1$.

In order to complete the proof of (2.18) and (2.19), the estimates (2.21), (2.22), (2.23), and (2.24) show that it is enough to prove that

$$\sum_{k=-\infty}^{+\infty} \int_{\mathbb{R}^m} |E_k(fb)|^2 |\triangle_k(b)|^2 dx \lesssim \|f\|_2^2. \tag{2.25}$$

The proof of this is more or less standard. We shall follow Christ [Ch] with very minor alterations. We make the following definition.

Definition 2.16. *We call a sequence of positive, measurable functions on $\mathbb{R}^m$,* $\omega = \{\omega_k\}_{k \in \mathbb{Z}}$, *Carleson if*

$$\|\omega\|_\mathcal{C} := \sup_{Q \in \mathcal{F}} \frac{1}{|Q|} \int_Q \left(\sum_{\{k \in \mathbb{Z}; 2^{-k} \le l(Q)\}} \omega_k(x)\, dx \right) < +\infty.$$

We need the following two (essentially) well known lemmas.

Lemma 2.17. *For every b in $BMO(\mathbb{R}^m)_{(n)}$, $\{|\triangle_k(b)|^2\}_k$ is Carleson, with norm not exceeding (a multiple of) $\|b\|^2_{BMO}$.*

Lemma 2.18. *Suppose $\omega = \{\omega_k\}_k$ is Carleson. Then, for any $1 < p < \infty$,*

$$\int_{\mathbb{R}^m} \sum_k |E_k f(x)|^p \omega_k(x)\, dx \lesssim \|\omega\|_\mathcal{C} \|f\|_p^p,$$

uniformly for $f \in L^p(\mathbb{R}^m)_{(n)}$.

Accepting these results, the inequalities (2.18) and (2.19) immediately follow and the proof of Theorem 2.14 is therefore complete. ∎

For the reader's convenience, we shall include the proofs of these lemmas.

Proof of Lemma 2.17. We need to estimate

$$\frac{1}{|Q|} \int_Q \left(\sum_{\{k \in \mathbb{Z}; 2^{-k} \le l(Q)\}} |\triangle_k(b)|^2 dx \right)$$

for an arbitrary, fixed, dyadic cube Q. Since $\triangle_k$ annihilates constants, we may suppose that b_Q, the integral mean of b over Q, is in fact zero. Write $b \in BMO(\mathbb{R}^m)_{(n)}$ as $b = b_0 + b_\infty$ with $b_0 = \chi_Q b$, and $b_\infty = \chi_{\mathbb{R}^m \setminus Q} b$. Then, by (2.23) and the John–Nirenberg inequality,

$$|Q|^{-1} \int_Q \sum_{2^{-k} \le l(Q)} |\triangle_k(b_0)(x)|^2 dx \le |Q|^{-1} \int_{\mathbb{R}^m} \sum_{k \in \mathbb{Z}} |\triangle_k(b_0)(x)|^2 dx$$

$$= |Q|^{-1} \|b_0\|_2^2 = |Q|^{-1} \int_Q |b|^2 dx = |Q|^{-1} \int_Q |b - b_Q|^2 dx \lesssim \|b\|^2_{BMO}.$$

Next, we want to estimate $|\triangle_k(b_\infty)(x)|$ for a fixed $x \in Q$ with $l(Q) \le 2^{-k}$. We note that there are unique dyadic cubes $Q' \in \mathcal{F}_k$ and $Q'' \in \mathcal{F}_{k+1}$ which contain x, and

37

these cubes must be contained in Q. Since $b_\infty \equiv 0$ on Q, we have

$$\triangle_k b_\infty(x) = \frac{1}{|Q''|} \int_{Q''} b_\infty \, dx - \frac{1}{|Q'|} \int_{Q'} b_\infty \, dx = 0,$$

and the proof is complete. ∎

Proof of Lemma 2.18. For an arbitrary $1 < p < \infty$, we have

$$\int_{\mathbb{R}^m} \sum_k |E_k f(x)|^p \omega_k(x) \, dx = p \int_0^\infty \sum_k \omega_k(\{x; |E_k f(x)| > \lambda\}) \lambda^{p-1} d\lambda \qquad (2.26)$$

For each fixed λ we choose a maximal collection $\{Q_j\}$ of pairwise disjoint dyadic cubes among the dyadic cubes Q such that

$$\left| \frac{1}{|Q|} \int_Q f(x) \, dx \right| > \lambda.$$

If we let $M f(x) := \sup_{k \in \mathbb{Z}} |E_k f(x)|$, then clearly $|\{x; |M f(x)| > \lambda\}| = \sum_j |Q_j|$. Furthermore, if $|Q|^{-1} \left| \int_Q f \, dx \right| > \lambda$ for some $Q \in \mathcal{F}_k$, then, by the maximality of the Q_j's, Q is contained in exactly one of the Q_j's. As a consequence,

$$\sum_k \omega_k(\{x; |E_k f(x)| > \lambda\}) \lesssim \sum_j \sum_{\{k; 2^{-k} \le l(Q_j)\}} \omega_k(Q_j) \lesssim \|\omega\|_C \sum_j |Q_j|$$
$$\lesssim \|\omega\|_C \, |\{x; |M f(x)| > \lambda\}|.$$

Inserting this in (2.26), the L^p−boundedness of maximal function gives

$$\int_{\mathbb{R}^m} \sum_k |E_k f(x)|^p \omega_k(x) \, dx \lesssim \|\omega\|_C \|M f\|_p^p \lesssim \|\omega\|_C \|f\|_p^p.$$

∎

The last result we prove in this section is that the just constructed Haar Clifford wavelets are an unconditional basis of $L^p(\mathbb{R}^m)_{(n)}$, for $1 < p < \infty$.

Theorem 2.19. For $1 < p < \infty$ and $f : \mathbb{R}^m \longrightarrow \mathbb{C}_{(n)}$ locally integrable, the following are equivalent:

(1) $f \in L^p(\mathbb{R}^m)_{(n)}$;

(2) $f = \sum_{Q \in \mathcal{F}} \sum_{i=1}^{2^m - 1} B(f, \Theta_{Q,i}^R) \Theta_{Q,i}^L$ with convergence in $L^p(\mathbb{R}^m)_{(n)}$;

(3) $A(x) := \left(\sum_{Q \in \mathcal{F}} \sum_{i=1}^{2^m - 1} |B(f, \Theta_{Q,i}^R) \Theta_{Q,i}^L(x)|^2 \right)^{1/2} \in L^p(\mathbb{R}^m)$;

(4) $A'(x) := \left(\sum_{Q \in \mathcal{F}} \sum_{i=1}^{2^m - 1} |B(f, \Theta_{Q,i}^R)|^2 |Q|^{-1} \chi_Q(x) \right)^{1/2} \in L^p(\mathbb{R}^m)$.

Moreover, if the above conditions are fulfilled, then

$$\|f\|_{L^p} \approx \|A\|_{L^p} \approx \|A'\|_{L^p}.$$

Also, similar results are valid for Θ^R's.

Proof. By (2.20) and Lemma 2.15, (2) is equivalent to $f = \sum_{k \in \mathbb{Z}} (E_{k+1}^L - E_k^L)f$ in $L^p(\mathbb{R}^m)_{(n)}$. It is not difficult to see that the sequence of bounded operators in L^p, $\{E_k^L\}_{k \in \mathbb{Z}}$, satisfies

$$E_k^L \longrightarrow \begin{cases} I, & \text{as } k \longrightarrow +\infty \\ \\ 0, & \text{as } k \longrightarrow -\infty, \end{cases}$$

in the strong operator norm. Therefore (1) $\Longleftrightarrow$ (2).

We consider next the equivalence (1) $\Longleftrightarrow$ (3) and introduce the operators

$$T_\omega(f)(x) := \sum_{Q \in \mathcal{F}} \sum_i \omega_{Q,i} B(f, \Theta_{Q,i}^R) \Theta_{Q,i}^L(x), \quad f \in L^2 \cap L^p,$$

where $\omega = \{\omega_{Q,i}\}$, with $Q \in \mathcal{F}$ and $i = 1, 2, ..., 2^m - 1$, is a sequence of ± 1. Clearly, for any such ω, T_ω is a bounded operator in $L^2(\mathbb{R}^m)_{(n)}$. We claim that in fact T_ω is also of weak-type $(1, 1)$. To prove this claim, for a given $f \in L^1 \cap L^2$ and $\lambda > 0$, we perform the Calderón-Zygmund decomposition for f at the level λ (cf. [St]). Hence, we can write $f = g + b$, where the "bad" b part is decomposed further as $b = \sum_j b_j$, where $\operatorname{supp} b_j \subseteq Q^j \in \mathcal{F}$, $\int_{Q^j} b_j = 0$ and $\sum_j |Q^j| \lesssim \lambda^{-1} \|f\|_{L^1}$. Consequently,

$$T_\omega(b_j)(x) = \sum_{Q \in \mathcal{F}} \sum_i \omega_{Q,i} \left(\int_{Q^j} b_j(y) \Theta_{Q,i}^R(y) \, dy \right) \Theta_{Q,i}^L(x).$$

Using the vanishing moment property and the precise localization of Θ's and b_j's, we see that this sum has only zero terms for $Q \neq Q^j$. In particular, $\operatorname{supp} T_\omega(b_j) \subseteq Q^j$ which, in turn, implies that

$$|\{x; \, |T_\omega(b)| > \lambda\}| \leq \sum_j |Q^j| \lesssim \lambda^{-1} \|f\|_{L^1},$$

and the claim is proved. The usual interpolation argument then yields the L^p-boundedness of T_ω for $1 < p \leq 2$.

The dual range is dealt with by a fairly standard duality argument, which we include for the sake of completeness. Let T_ω^* be the adjoint of T_ω with respect to the form $B(\cdot, \cdot)$, i.e. T_ω^* is the unique continuous right-Clifford-linear operator in L^2 for which

$$B(T_\omega f, g) = B(f, T_\omega^* g), \quad f, g \in L^2.$$

As before, we can display the kernel of T_ω^* in terms of Θ's, and using the same argument we get that T_ω^* is L^p-bounded for $1 < p < 2$. Now, if $2 < p < \infty$ and q is its conjugate exponent, we have

$$|\langle T_\omega f, g \rangle| = |B(T_\omega f, b^{-1} g)| = |B(f, T_\omega^*(b^{-1} g))| = |\langle f, b T_\omega^*(b^{-1} g) \rangle|$$

$$\lesssim \|f\|_{L^p} \|b T_\omega^*(b^{-1} g)\|_{L^q} \lesssim \|f\|_{L^p} \|g\|_{L^q},$$

since $1 < q < 2$ and $b^{-1} \in L^\infty$. Hence $\|T_\omega f\|_{L^p} \lesssim \|f\|_{L^p}$, and since $T_\omega^2 = I$, we get the equivalence $\|T_\omega f\|_{L^p} \approx \|f\|_{L^p}$, uniformly in $\omega \in \{-1, +1\}^{\mathcal{F} \times \{1,2,\dots,2^m-1\}}$. Finally, we integrate this equivalence against the measure $d\mu(\omega)$ on $\{-1, +1\}^{\mathcal{F} \times \{1,2,\dots,2^m-1\}}$ given by $d\mu := \otimes_{Q,i} d\nu$, where $d\nu$ is the probability measure on $\{-1, +1\}$ taking the value $1/2$ on $\{\pm 1\}$, and then use Khintchine's lemma asserting that on the measure space $(\{-1, +1\}^{\mathcal{F} \times \{1,2,\dots,2^m-1\}}, d\mu)$ any L^p norm is equivalent to the L^2 norm (see also [**Me**]). The conclusion follows, and the proof of (1) $\iff$ (3) is complete.

Obviously, (4) implies (3). To see the converse implication, we first note that $|B(f, \Theta_{Q,i}^R) \Theta_{Q,i}^L(x)| \approx |B(f, \Theta_{Q,i}^R)| |\Theta_{Q,i}^L(x)|$, since the nonzero values of $\Theta_{Q,i}^L$ are in the Clifford group of $\mathbb{R}_{(n)}$. Thus, any dyadic cube Q has a children Q' on which $A(x) \approx A'(x)$. Finally, one can use Lebesgue's differentiation theorem to conclude that in fact $A(x) \approx A'(x)$ for a.e. $x \in \mathbb{R}^m$. ∎

Remarks.

(A) It should be noted that a Littlewood-Paley type estimate of the form

$$\left\| \left(\sum_k \left| E_{k+1}^L f - E_k^L f \right|^2 \right)^{1/2} \right\|_{L^2} \lesssim \|f\|_{L^2}$$

is valid under more general circumstances, e.g. as part of a Clifford-martingale theory as developed in [**GLQ**].

(B) In the construction of the Haar Clifford wavelets the family of all dyadic cubes in $\mathbb{R}^m$ can be replaced by a more general system $\mathcal{F} = \cup_k \mathcal{F}_k$ subject to the following set of conditions (all constants involved being independent of k):

(1) For each $k \in \mathbb{Z}$, $\mathcal{F}_k = \{Q_{k,j}\}_j$ is a countable partition of $\mathbb{R}^m$ consisting of measurable sets of finite Lebesgue measure which satisfy $(\operatorname{diam} Q)^m \leq \operatorname{const} |Q|$ for all $Q \in \mathcal{F}_k$.

(2) If $Q \in \mathcal{F}_k$ and $Q' \in \mathcal{F}_{k+1}$ are not disjoint, then necessarily $Q' \subsetneq Q$ and $1 < \operatorname{const} \leq |Q|/|Q'| \leq \operatorname{const} < +\infty$.

(3) $b(x)$ is a L^∞-function which takes on values in $\mathbb{R}^{n+1}$ and for which (2.14) holds for any $Q \in \mathcal{F}$.

This allows us to handle the case when b is a *pseudo-accretive* function, i.e. b is a L^∞-function for which there exist $\delta > 0$ and $C > 0$ such that for any $x \in \mathbb{R}^n$ and any $r > 0$, one can find a cube $Q \subset B_r(x)$ with $r \leq C\, l(Q)$ for which

$$\left| \frac{1}{|Q|} \int_Q b(y)\, dy \right| \geq \delta. \tag{2.27}$$

Roughly speaking, this means that for any cube Q there exists another cube Q' at about the same place and of about the same size for which (2.27) holds. The idea (due to David) is to set up a class of "cubes" $\mathcal{F}$ by suitably correcting each dyadic cube such that, in the end, for any "cube" $Q \in \mathcal{F}$, (2.27) is fulfilled (eventually for some smaller δ; see [CJS]).

(C) The construction also works on spaces of homogeneous type (see e.g. [CW]).

Chapter 3

The L^2 Boundedness of Clifford Algebra Valued Singular Integral Operators

The L^2−boundedness of the higher dimensional Cauchy integral operator on a Lipschitz hypersurface can be reduced, by means of the rotation method of Calderón and Zygmund, to the special case of a Lipschitz curve and in that case one is in a position to use the celebrated result of Coifman, McIntosh and Meyer [CMM]. However, it is natural to try to prove this without relying on the rotation method, working directly on the surface and the first results in this direction are due to Coifman, Murray and McIntosh ([Mu], [Mc]).

To explain our approach, we first need to introduce some notation. We shall work in the Euclidean space $\mathbb{R}^{n+1}$ canonically embedded in the Clifford algebra $\mathbb{R}_{(n)}$. Let $g : \mathbb{R}^n \longrightarrow \mathbb{R}$ be a Lipschitz continuous function, i.e. for some $M > 0$ one has $|g(x) - g(y)| \leq M|x - y|$, for all $x, y \in \mathbb{R}^n$, and denote by Σ its graph:

$$\Sigma := \{X = (g(x), x) \,;\, x \in \mathbb{R}^n\} \subseteq \mathbb{R}^{n+1} \hookrightarrow \mathbb{R}_{(n)}.$$

The exterior normal $N(x) := (-1, \nabla g(x))$ is then a well-defined Clifford vector for almost all $x \in \mathbb{R}^n$, and $\operatorname{Re} N(x) = -1$, $|N(x)| \approx 1$. Also, $n := N/|N|$ is the unit normal on Σ.

Formally, in local coordinates, the Cauchy singular integral operators on Σ are defined by

$$C^L f(x) := \lim_{\delta \to 0} \frac{1}{\sigma_n} \int_{\mathbb{R}^n} f(y) N(y) \frac{\overline{z(y) - z(x) - \delta}}{|z(y) - z(x) - \delta|^{n+1}} \, dy,$$

and

$$C^R f(x) := \lim_{\delta \to 0} \frac{1}{\sigma_n} \int_{\mathbb{R}^n} \frac{\overline{z(y) - z(x) - \delta}}{|z(y) - z(x) - \delta|^{n+1}} N(y) f(y) \, dy,$$

where σ_n is the area of the unit sphere in $\mathbb{R}^{n+1}$, $z(x) := (g(x), x) \in \mathbb{R}^{n+1} \subset \mathbb{R}_{(n)}$, and f is a $\mathbb{C}_{(n)}$-valued function on $\mathbb{R}^n$. Note that the above integrands are to be understood in the sense of multiplication in $\mathbb{C}_{(n)}$.

42

We shall follow the second part of [CJS] and prove the L^2-boundedness of C^L and C^R. The idea is to adapt the standard inner product for $L^2(\mathbb{R}^n)$ to the geometry of Σ by considering the Clifford bilinear form $\langle \cdot, \cdot \rangle_\Sigma$

$$\langle f_1, f_2 \rangle_\Sigma := \int_{\mathbb{R}^n} f_1(x) N(x) f_2(x) \, dx$$

defined for $\mathbb{C}_{(n)}$-valued functions f_1 and f_2. Considering the Haar Clifford wavelets constructed in the previous chapter which correspond to the 1-accretive, L^∞-function $N(x)$ and expressing the Cauchy integral as a matrix operator with respect to this basis, the boundedness will be a consequence of (a version of) the ordinary Schur's lemma.

It is interesting to note that, in the commutative case i.e. for $n = 1$, essentially the same reasoning yields even more, namely David's theorem ([Dav1]) concerning the boundedness of the Cauchy integral on the so called chord-arc curves (cf. also [CJS]).

In §3.1 we discuss in detail the boundedness of the Cauchy singular integral operator while in section §3.3 we briefly indicate how these techniques may be used to prove the Clifford algebra version of the $T(b)$ theorem.

§3.1 THE HIGHER DIMENSIONAL CAUCHY INTEGRAL

As much as possible, we shall keep the notations introduced so far. For a fixed $\delta \neq 0$ and for $f \in L^2(\mathbb{R}^n)_{(n)}$, we define the operators C_δ^L and C_δ^R by

$$C_\delta^L f(x) := \frac{1}{\sigma_n} \int_{\mathbb{R}^n} f(y) N(y) \frac{\overline{z(y) - z(x) - \delta}}{|z(y) - z(x) - \delta|^{n+1}} \, dy, \quad x \in \mathbb{R}^n,$$

and

$$C_\delta^R f(x) := \frac{1}{\sigma_n} \int_{\mathbb{R}^n} \frac{\overline{z(y) - z(x) + \delta}}{|z(y) - z(x) + \delta|^{n+1}} N(y) f(y) \, dy, \quad x \in \mathbb{R}^n,$$

respectively.

Let $\{\Theta_{Q,i}^L, \Theta_{Q,i}^R\}_{Q,i}$ be the Haar Clifford wavelets associated to the Clifford bilinear form $\langle \cdot, \cdot \rangle_\Sigma$. The main estimates in the proof of the L^2-boundedness of these operators are contained in the next theorem.

43

Theorem 3.1. *Let $0 < \epsilon < \frac{1}{n}$ and, for any dyadic cube of $\mathbb{R}^n$, let $w(Q) := |Q|^{\frac{1}{2}-\epsilon}$. Then we have*

$$\sup_{Q,i} w(Q)^{-1} \sum_{Q'} \sum_{j=1}^{2^n-1} w(Q') \left| \langle C_\delta^L \Theta_{Q,i}^L, \Theta_{Q',j}^R \rangle_\Sigma \right| \leq C < +\infty, \qquad (3.1)$$

$$\sup_{Q',j} w(Q')^{-1} \sum_{Q} \sum_{i=1}^{2^n-1} w(Q) \left| \langle C_\delta^L \Theta_{Q,i}^L, \Theta_{Q',j}^R \rangle_\Sigma \right| \leq C < +\infty, \qquad (3.2)$$

with C independent of δ. Similar estimates are also valid for C_δ^R.

We shall postpone the proof of this for a moment and first derive some of its consequences. First recall the following version of Schur's test.

Lemma 3.2. *Assume that the rows and the columns of an infinite matrix $A = (a_{ij})_{i,j}$ satisfy*

$$w_i^{-1} \sum_j |a_{ij}| w_j \leq C < +\infty \quad \text{for each } i,$$

and

$$w_j^{-1} \sum_i |a_{ij}| w_i \leq C < +\infty \quad \text{for each } j,$$

for some constant $C > 0$ and some positive numbers $(w_j)_j$. Then A defines a bounded operator on $\ell^2(\mathbb{N})$ with norm $\leq C$.

A combination of this and Theorem 3.1 immediately yields the following.

Corollary 3.3. *The operators C_δ^L and C_δ^R are bounded on $L^2(\mathbb{R}^n)_{(n)}$, with bounds independent of δ.*

Using the Poisson-like decay of the kernel of $C_\delta^L - C_{-\delta}^L$ one can readily see that $(C_\delta^L - C_{-\delta}^L)f \longrightarrow f$ in L^p as $\delta \longrightarrow 0$, for any $f \in L^2(\mathbb{R}^n)_{(n)}$. Moreover, since $C_{\delta'}^L(C_\delta^L - C_{-\delta}^L)f = C_{\delta'+\delta}^L f \longrightarrow C_\delta^L f$ in L^2 as $\delta' \longrightarrow 0$, we see that the limit $\lim_{\delta \to 0} C_\delta^L f$ exists for any $f \in L^2(\mathbb{R}^n)_{(n)}$. In fact, the next theorem (whose proof will be completed in the next chapter) shows that we can be very precise about the limit operator.

For a fixed, $\delta > 0$, we introduce *the truncated Hilbert transforms* H_δ^L and H_δ^R by

$$H_\delta^L f(X) := \frac{2}{\sigma_n} \int_{\substack{|X-Y| \geq \delta \\ Y \in \Sigma}} f(Y) n(Y) \frac{\overline{Y-X}}{|Y-X|^{n+1}} \, dS(Y), \quad X \in \Sigma,$$

and

$$H_\delta^R f(X) := \frac{2}{\sigma_n} \int_{\substack{|X-Y|\geq\delta \\ Y\in\Sigma}} \frac{\overline{Y-X}}{|Y-X|^{n+1}} n(Y) f(Y)\, dS(Y), \quad X \in \Sigma,$$

where dS is the canonical surface measure on Σ.

Simple, direct computations involving the Cauchy kernel $E(X)$ show that, for $\delta > 0$,

$$\left| C_\delta^L f(x) - H_\delta^L (f \circ z^{-1})(z(x)) \right| \lesssim f^*(x), \quad x \in \mathbb{R}^n,$$

uniformly in δ. Here $*$ is the standard Hardy-Littlewood maximal operator. A similar estimate holds for the difference between C_δ^R and H_δ^R. In particular, $\{H_\delta^L\}_\delta$ and $\{H_\delta^R\}_\delta$ are bounded in $L^2(\mathbb{R}^n)_{(n)}$ with bounds independent of δ. As a consequence, the Cauchy kernel is a Calderón-Zygmund kernel; see the Appendix of [CMM]. With this and some standard arguments from the theory of the Calderón-Zygmund operators, as presented in e.g. [Me], we get the following.

Theorem 3.4. *The Cauchy singular integral operators, or Hilbert transforms on Σ, H^L and H^R defined for any $f \in L^p(\Sigma, dS)_{(n)}$ and almost all $X \in \Sigma$ by*

$$H^L f(X) := \lim_{\delta\to+0} H_\delta^L f(X) \quad \text{and} \quad H^R f(X) := \lim_{\delta\to+0} H_\delta^R f(X)$$

are well-defined and bounded on $L^2(\Sigma, dS)_{(n)}$.

Moreover,

$$\lim_{\delta\to\pm0} C_\delta^L f(x) = \frac{1}{2}(\pm I + H^L) F(X)$$

and

$$\lim_{\delta\to\mp0} C_\delta^R f(x) = \frac{1}{2}(\pm I + H^R) F(X),$$

at almost any $x \in \mathbb{R}^n$, where $F := f \circ z^{-1}$, $X := z(x) \in \Sigma$.

The rest of this section is devoted to presenting the proof of Theorem 3.1. First we shall establish some estimates on the "size" of the image of a Haar Clifford wavelet under the action of the Cauchy integral. We emphasize that all the constants involved are independent of δ.

Lemma 3.5. *For any dyadic cube Q, and any $i = 1, 2, ..., 2^n - 1$, we have :*

$$\left|C_\delta^L \Theta_{Q,i}^L(x)\right| \lesssim l(Q)|Q|^{1/2}|x - x_Q|^{-(n+1)}, \text{ if } x \notin 2Q, \tag{3.3}$$

and

$$\left|C_\delta^L \Theta_{Q,i}^L(x)\right| \lesssim |Q|^{-1/2} \log \frac{c_n l(Q)}{dist\,(x, \partial_0 Q)}, \text{ if } x \in 2Q, \tag{3.4}$$

where $\partial_0 Q := \cup_{j=1}^{2^n} \partial Q^j$, with $\{Q^j\}_j$ the "children" of Q, and x_Q is the "leftmost corner" of Q, i.e. for $Q = Q_{k,v}$, $x_Q := 2^{-k}v$. Here c_n denotes a constant which depends only on n. Analogous estimates hold for $C_\delta^R \Theta_{Q,j}^R$ as well.

Proof. Inequality (3.3) is a consequence of the fact that Θ's have vanishing moment

$$|C_\delta^L \Theta_{Q,i}^L(x)| =$$
$$\frac{1}{\sigma_n} \left| \int_{\mathbb{R}^n} \Theta_{Q,i}^L(y) N(y) \left(\frac{\overline{z(y) - z(x) - \delta}}{|z(y) - z(x) - \delta|^{n+1}} - \frac{\overline{z(x_Q) - z(x) - \delta}}{|z(x_Q) - z(x) - \delta|^{n+1}} \right) dy \right|.$$

Hence, (3.3) follows by using the mean-value theorem to estimate the the expression inside the parentheses and by noting that for $x \notin 2Q$, $dist\,(x, Q) \gtrsim |x - x_Q|$.

Next we consider (3.4). If $x \in 2Q,'$ we have

$$|C_\delta^L \Theta_{Q,i}^L(x)| \approx \left| \int_{\mathbb{R}^n} \Theta_{Q,i}^L(y) N(y) \frac{\overline{z(y) - z(x) - \delta}}{|z(y) - z(x) - \delta|^{n+1}} dy \right|$$
$$\lesssim |Q|^{-1/2} \sum_{j=1}^{2^n} \left| \int_{Q^j} N(y) \frac{\overline{z(y) - z(x) - \delta}}{|z(y) - z(x) - \delta|^{n+1}} dy \right|.$$

Let $d := dist\,(x, \partial_0 Q) \le dist\,(x, \partial Q^j)$, for all j. If $x \in 2Q \setminus Q$, we majorize each integral in the above sum by

$$\int_{d < |x-y| \le d + diam(Q)} |x - y|^{-n}\, dy \lesssim \log\left(2n^{1/2}l(Q)/d\right),$$

which gives a bound of the right order. Assume now that $x \in Q^j$ for some j. This time we split each integral as

$$\int_{Q^j} = \int_{|X-Y|>d} + \int_{|X-Y|<d} =: I + II,$$

46

where $X := z(x)$, $Y := z(y)$. Since $|x - y| \approx |X - Y|$, we get that I has the right size in the same way as before. As for II, we note that if Ω_- is the domain in $\mathbb{R}^{n+1}$ below the surface Σ then, by the monogenicity of the Cauchy kernel,

$$II = - \int_{\substack{|X-Y|=d \\ Y \in \Omega_-}} \frac{Y - X}{|Y - X|} \frac{\overline{Y - X - \delta}}{|Y - X - \delta|^{n+1}} \, d\sigma_d(Y),$$

where $d\sigma_d$ is the standard surface measure on the sphere of radius d centered at the origin of $\mathbb{R}^{n+1}$. Hence,

$$|II| \lesssim \int_{\substack{|X-Y|=d \\ Y \in \mathbb{R}^{n+1}}} \frac{1}{|X - Y|^n} \, d\sigma_d(Y) = \sigma_n,$$

which completes the proof of the lemma. ∎

Lemma 3.6. *For $f \in L^\infty_{\text{comp}}(\mathbb{R}^n)_{(n)}$ satisfying the cancellation condition $\langle f, 1 \rangle_\Sigma = 0$ we have $\langle C^L_\delta f, 1 \rangle_\Sigma = 0$. Also, if $\langle 1, f \rangle_\Sigma = 0$, then $\langle 1, C^R_\delta f \rangle_\Sigma = 0$.*

Proof. These are both simple consequences of Cauchy's vanishing theorem (see Chapter 1) applied to the domains Ω_-, Ω_+ located below and above Σ, respectively, and to the functions $C^L_\delta f$, $C^R_\delta f$ which are right monogenic in a neighborhood of $\overline{\Omega_-}$ and left monogenic in a neighborhood of $\overline{\Omega_+}$, respectively. That this works is guaranteed by the good decay of the functions at infinity (as in the proof of the previous lemma, the vanishing moment of f actually improves the decay of $C^L_\delta f$ and $C^R_\delta f$). ∎

Our next lemma shows that C^L_δ and C^R_δ are essentially adjoint to each other with respect to the form $\langle \cdot, \cdot \rangle_\Sigma$.

Lemma 3.7. *We have $\langle C^L_\delta f, f' \rangle_\Sigma = -\langle f, C^R_\delta f' \rangle_\Sigma$, for any $f, f' \in L^\infty_{\text{comp}}(\mathbb{R}^n)_{(n)}$. In particular, $\langle C^L_\delta \Theta^L_{Q,i}, \Theta^R_{Q',j} \rangle_\Sigma = -\langle \Theta^L_{Q,i}, C^R_\delta \Theta^R_{Q',j} \rangle_\Sigma$.*

Proof. This is immediate by Fubini's theorem since the double integral

$$\int_{\mathbb{R}^n} \int_{\mathbb{R}^n} f(y) N(y) \frac{\overline{z(y) - z(x) - \delta}}{|z(y) - z(x) - \delta|^{n+1}} N(x) f'(x) \, dx \, dy$$

is absolutely convergent. ∎

Let us now consider a dyadic cube $Q_{k,v}$. We let Φ be the linear rescaling mapping of $Q_{k,v}$ onto the standard unit cube in $\mathbb{R}^n$, i.e. $\Phi(y) := w := 2^k y - v$. For an arbitrary dyadic cube Q we set $Q^* := \Phi(Q)$. We have that Q^* is still dyadic as long as $l(Q) \leq 2^{-k}$. Furthermore, if $z^*(w) := 2^k z(2^{-k}(w+v))$, for $w \in \mathbb{R}^n$, one can readily see that z^* is a bi-Lipschitz application with constants comparable to those of the initial z. Let Σ^* be the graph of z^*. As a general rule, we agree that the superscript $*$ is used to label objects related to Σ^* which are analogous to those we have constructed in connection with Σ.

Direct calculation shows that

$$\Theta^{L*}_{Q^*,i}(w) = 2^{-kn/2}\Theta^L_{Q,i}(y), \qquad C^{L*}_\delta \Theta^{L*}_{Q^*,i}(w) = 2^{-kn/2}C^L_{2^{-k}\delta}\Theta^L_{Q,i}(y),$$

and

$$\langle C^{L*}_{2^k\delta}\Theta^{L*}_{Q^*,i}, \Theta^{R*}_{Q'^*,j}\rangle_{\Sigma^*} = \langle C^L_\delta \Theta^L_{Q,i}, \Theta^R_{Q',j}\rangle_\Sigma.$$

In addition, $w^*(Q^*) = 2^{nk(1/2-\epsilon)}w(Q)$, so that

$$w^*(Q'^*)^{-1} \sum_{\{Q\in\mathcal{F};l(Q^*)\leq l(Q'^*)\}} \sum_{i=1}^{2^n-1} w^*(Q^*)\left|\langle C^{L*}_\delta \Theta^{L*}_{Q^*,i}, \Theta^{R*}_{Q'^*,j}\rangle_{\Sigma^*}\right|$$

$$= w(Q')^{-1} \sum_{\{Q\in\mathcal{F};l(Q)\leq l(Q')\}} \sum_{i=1}^{2^n-1} w(Q)\left|\langle C^L_{l(Q')\delta}\Theta^L_{Q,i}, \Theta^R_{Q',j}\rangle_\Sigma\right|,$$

for each $Q' \in \mathcal{F}$ and $j = 1, 2, ..., 2^n - 1$. Consequently, in order to finish the proof of (3.1), it is enough to prove that, for any j,

$$\sum_{\{Q\in\mathcal{F};1\leq l(Q)\}} \sum_{i=1}^{2^n-1} w(Q)\left|\langle C^L_\delta \Theta^L_{Q,i}, \Theta^R_{[0,1]^n,j}\rangle_\Sigma\right| \leq C < +\infty,$$

and that, for any i,

$$\sum_{\{Q\in\mathcal{F};l(Q)\leq 1\}} \sum_{j=1}^{2^n-1} w(Q)\left|\langle \Theta^L_{[0,1]^n,i}, C^R_\delta \Theta^R_{Q,j}\rangle_\Sigma\right| \leq C < +\infty,$$

where C may depend on $\|\nabla g\|_{L^\infty}$ but is independent of δ and i, j. If we combine these estimates with the similar ones needed to complete the proof of (3.2), we see

48

that we actually must prove that for all i, j,

$$\sum_{Q \in \mathcal{F}} w(Q) \left| \langle C_\delta^L \Theta_{Q,i}^L, \Theta_{[0,1]^n,j}^R \rangle_\Sigma \right| \le C < +\infty \tag{3.5}$$

and

$$\sum_{Q \in \mathcal{F}} w(Q) \left| \langle \Theta_{[0,1]^n,i}^L, C_\delta^R \Theta_{Q,j}^R \rangle_\Sigma \right| \le C < +\infty, \tag{3.6}$$

with C possibly depending on $\| \nabla g \|_{L^\infty}$ but not on δ.

The proofs of (3.5) and (3.6) are virtually the same, and we confine ourselves to showing e.g. (3.5). We need to discuss several cases.

Case I. $l(Q)$ "large" and Q "clearly" disjoint from the standard unit cube: $l(Q) \ge 1$ and $2Q \cap [0,1]^n = \varnothing$.

Using (3.3) and that $|x - x_Q| \gtrsim |x_Q|$, we get

$$w(Q) \left| \langle C_\delta^L \Theta_{Q,i}^L, \Theta_{[0,1]^n,j}^R \rangle_\Sigma \right| \lesssim |Q|^{1/2 - \epsilon} l(Q) |Q|^{1/2} |x_Q|^{-(n+1)}. \tag{3.7}$$

Now, if $Q = Q_{k,v}$, our hypotheses imply that $v \ne 0$ so that, as $\epsilon > 0$, the right-hand side of (3.7) is majorized by

$$\sum_{k=0}^{-\infty} \sum_{\substack{v \ne 0 \\ v \in \mathbb{Z}^n}} 2^{-kn(1/2-\epsilon)} 2^{-k} 2^{-kn/2} 2^{k(n+1)} |v|^{-(n+1)}$$

$$= \left(\sum_{k=0}^{-\infty} 2^{kn\epsilon} \right) \left(\sum_{\substack{v \ne 0 \\ v \in \mathbb{Z}^n}} |v|^{-(n+1)} \right) < +\infty,$$

which proves that the corresponding piece in (3.5) satisfies the right estimate.

Case II. $l(Q)$ is "large", i.e. $l(Q) \ge 1$, but $[0,1]^n \cap 2Q \ne \varnothing$.

Note that $[0,1]^n \cap 2Q \ne \varnothing$ implies that there exists a fixed nonnegative integer M_0 (3^n will do), such that for any k, $\mathcal{F}_k$ contributes with at most M_0 dyadic cubes to this case. Now, by Lemma 3.7, (3.3) and (3.4),

$$w(Q) \left| \langle C_\delta^L \Theta_{Q,i}^L, \Theta_{[0,1]^n,j}^R \rangle_\Sigma \right| = w(Q) \left| \langle \Theta_{Q,i}^L, C_\delta^R \Theta_{[0,1]^n,j}^R \rangle_\Sigma \right|$$

49

$$\lesssim |Q|^{1/2-\epsilon}|Q|^{-1/2}\left(\int_{Q\cap2[0,1]^n}|C_\delta^R\Theta_{[0,1]^n,j}^R(x)|dx+\int_{Q\setminus2[0,1]^n}|C_\delta^R\Theta_{[0,1]^n,j}^R(x)|dx\right)$$

$$\lesssim |Q|^{-\epsilon}\left(\int_{2[0,1]^n}\log\frac{c_n}{\operatorname{dist}(x,\partial_0[0,1]^n)}\,dx+\int_{\mathbb{R}^n\setminus2[0,1]^n}|x|^{-(n+1)}dx\right)\approx|Q|^{-\epsilon}.$$

Hence,

$$w(Q)\left|\langle C_\delta^L\Theta_{Q,i}^L,\Theta_{[0,1]^n,j}^R\rangle_\Sigma\right|\lesssim|\dot Q|^{-\epsilon}.$$

Using this, the part of (3.5) corresponding to this case can be estimated by

$$\sum_{k=0}^{-\infty}2^{kn\epsilon}\sum_{v\in\text{finite set}}1\approx\sum_{k=0}^{-\infty}2^{kn\epsilon}<+\infty,$$

as desired.

Case III. $l(Q)$ *"small" and Q "clearly separated" from the standard unit cube:*
$l(Q)<1$ *and* $2Q\cap[0,1]^n=\varnothing.$

This time we have, using (3.3) again,

$$w(Q)\left|\langle C_\delta^L\Theta_{Q,i}^L,\Theta_{[0,1]^n,j}^R\rangle_\Sigma\right|\lesssim|Q|^{1/2-\epsilon}\int_{[0,1]^n}|C_\delta^L\Theta_{Q,i}^L(x)|dx$$

$$\lesssim l(Q)|Q|^{1-\epsilon}|x_Q|^{-(n+1)}.$$

Assume that $Q=Q_{k,v}$. Since $2Q\cap[0,1]^n=\varnothing$, we must have $|v|\gtrsim2^k$. The appropriate part of the sum in (3.5) is therefore majorized by

$$\sum_{k=0}^{+\infty}2^{-k}2^{-kn(1-\epsilon)}2^{k(n+1)}\sum_{|v|\gtrsim2^k}|v|^{-(n+1)}\approx\sum_{k=0}^{+\infty}2^{k(n\epsilon-1)}<+\infty,$$

provided $0<\epsilon<1/n$.

There remains

Case IV. $l(Q)<1$ *and* $Q\subset[-3,3]^n.$

Let us first analyze the situation when $Q\subset[-3,0]\times[-3,3]^{n-1}$.

To estimate $|\langle C_\delta^L\Theta_{Q,i}^L,\Theta_{[0,1]^n,j}^R\rangle_\Sigma|$, note that since $\Theta_{[0,1]^n,j}^R$ is a linear combination of the characteristic functions of the "children" of the standard unit cube, it is enough to control

$$\left|\langle C_\delta^L\Theta_{Q,i}^L,\chi_{Q'}\rangle_\Sigma\right|\lesssim\int_{Q'}\left|C_\delta^L\Theta_{Q,i}^L(x)\right|\,dx,\qquad(3.8)$$

for an arbitrary fixed "child" Q' of $[0,1]^n$.

The first possibility is that the boundary of Q has no common points with the hyper-plane $\{x^1 = 0\}$. Then we may use (3.3) in (3.8) combined with the fact that $|x - x_Q| \gtrsim |x_Q^1|$ for $x \in [0,1]^n$ to dominate the integral by a multiple of

$$l(Q)|Q|^{1/2} \int_{|x_Q^1|}^{\infty} r^{-(n+1)} r^{n-1} dr = l(Q)|Q|^{1/2}|x_Q^1|^{-1}.$$

Hence, the contribution to the sum in (3.5) from this part has the upper bound

$$\sum_{k=0}^{+\infty} 2^{-kn(1/2-\epsilon)} 2^{-k} 2^{-kn/2} 2^k \sum_{\substack{-3\cdot 2^k \le v_i \le 3\cdot 2^k - 1 \\ i=2,\dots,n}} \sum_{v_1=1}^{-2^{k+1}} |v_1|^{-1} \approx \sum_{k=0}^{+\infty} k 2^{-k(1-n\epsilon)} < +\infty,$$

for $1/n > \epsilon > 0$.

The second possibility is that Q is adjacent to the hyper-plane $\{x^1 = 0\}$. Let us write $Q' = Q_1' \cup Q_2'$ with $Q_1' = 2Q \cap Q'$ and $Q_2' = Q' \setminus Q_1$. On the Q_2' part we still use (3.3) to majorize the integrand in (3.8). This and the fact that $|x - x_Q| \ge l(Q)$ show that

$$\int_{Q_2'} \left| C_\delta^L \Theta_{Q,i}^L(x) \right| dx \lesssim l(Q)|Q|^{1/2} \int_{Q_2'} |x - x_Q|^{-(n+1)} dx$$

$$\lesssim l(Q)|Q|^{1/2} \int_{l(Q)}^{+\infty} r^{-n-1} r^{n-1} dr \approx |Q|^{1/2}.$$

For each fixed $k \ge 0$ there are $\mathcal{O}(2^{k(n-1)})$ dyadic cubes which fit into this case, and

$$\sum_{k=0}^{+\infty} 2^{-kn(1/2-\epsilon)} 2^{-kn/2} 2^{k(n-1)} = \sum_{k=0}^{+\infty} 2^{-k(1-n\epsilon)} < +\infty,$$

for $1/n > \epsilon > 0$. Hence, we conclude that this part of the sum in (3.5) satisfies the right estimate as well.

As for the Q_1' part in (3.8), using (3.4) and $\operatorname{dist}(x, \partial_0 Q) \ge x^1 \ge 0$, we find

$$\int_{Q_1'} \left| C_\delta^L \Theta_{Q,i}^L(x) \right| dx \lesssim |Q|^{-1/2} \int_{Q_1'} \log \frac{c_n}{\operatorname{dist}(x, \partial_0 Q)} dx \lesssim |Q|^{-1/2} \int_{Q_1'} \log \frac{c_n}{x^1} dx$$

$$\lesssim |Q|^{-1/2} l(Q)^{n-1} \int_0^{l(Q)} \log \frac{c_n}{x^1} dx^1 = |Q|^{-1/2} \mathcal{O}(l(Q)^{n-1+a}),$$

since, for any $\alpha \in (0,1)$, $|t(\log(1/t)+1)| \lesssim t^{\alpha}$ uniformly for $t \in (0,1)$. Choose α such that $n\epsilon < \alpha < 1$. Then

$$\sum_{k=0}^{+\infty} 2^{-kn(1/2-\epsilon)} 2^{kn/2} 2^{-k(n-1+\alpha)} 2^{k(n-1)} = \sum_{k=0}^{+\infty} 2^{-k(\alpha-n\epsilon)} < +\infty.$$

We have now finished the proof when $Q \subset [-3,0] \times [-3,3]^{n-1}$. Next, we use the invariance of the boundedness of

$$\sum_Q w(Q) \left| \langle C_\delta^L \Theta_{Q,i}^L, \chi_{Q'} \rangle_\Sigma \right|$$

under translations, permutations of coordinates, and symmetries with respect to the coordinate axes. This allows us to reduce the problem to the one we just finished, whenever Q and Q' have disjoint interiors. In fact, the only remaining possibility we need to check in Case IV is when $Q \subseteq Q' = [0,1/2]^n$. On account of Lemma 3.6 we have

$$\left| \langle C_\delta^L \Theta_{Q,i}^L, \chi_{[0,1/2]^n} \rangle_\Sigma \right| = \left| \langle C_\delta^L \Theta_{Q,i}^L, \chi_{\mathbb{R}^n \setminus [0,1/2]^n} \rangle_\Sigma \right|$$
$$\lesssim \left| \langle C_\delta^L \Theta_{Q,i}^L, \chi_{\mathbb{R}^n \setminus 2[0,1/2]^n} \rangle_\Sigma \right| + \left| \langle C_\delta^L \Theta_{Q,i}^L, \chi_{2[0,1/2]^n \setminus [0,1/2]^n} \rangle_\Sigma \right| =: I + II. \quad (3.9)$$

Now II can be estimated by decomposing $2[0,1/2]^n \setminus [0,1/2]^n$ as an union of finitely many cubes which are disjoint from Q and for which we can apply the argument described in the preceding paragraph. To estimate I in (3.9), let B_R be the ball of radius R and centered at the origin of $\mathbb{R}^n$. We have that

$$\left| \langle C_\delta^L \Theta_{Q,i}^L, \chi_{\mathbb{R}^n \setminus 2[0,1/2]^n} \rangle_\Sigma \right| = \lim_{R \to \infty} \left| \langle C_\delta^L \Theta_{Q,i}^L, \chi_{B_R \setminus 2[0,1/2]^n} \rangle_\Sigma \right|$$
$$= \lim_{R \to \infty} \left| \langle \Theta_{Q,i}^L, C_\delta^R \chi_{B_R \setminus 2[0,1/2]^n} \rangle_\Sigma \right|$$
$$= \lim_{R \to \infty} \left| \langle \Theta_{Q,i}^L, C_\delta^R \chi_{B_R \setminus 2[0,1/2]^n} - C_\delta^R \chi_{B_R \setminus 2[0,1/2]^n}(x_Q) \rangle_\Sigma \right|.$$

Since $|\frac{\partial}{\partial x_j} C_\delta^R \chi_{B_R \setminus 2[0,1/2]^n}(x)| \lesssim 1$ for $x \in [0,1/2]^n$, $j = 1,2,...,n$, uniformly in R, we may bound the last limit from above by $|Q|^{1/2} l(Q)$. Each $\mathcal{F}_k$ contains $\mathcal{O}(2^{kn})$ dyadic cubes inside $[0,1/2]^n$ and

$$\sum_{k=0}^{+\infty} 2^{-kn(1/2-\epsilon)} 2^{-kn/2} 2^{-k} 2^{kn} = \sum_{k=0}^{+\infty} 2^{-k(1-n\epsilon)} < +\infty,$$

52

for $1/n > \epsilon > 0$. This finishes Case IV. The proof of (3.5) is therefore complete, hence so is the proof of Theorem 3.1. ∎

Remark. A more careful account of the way the Lipschitz constant $\| \nabla g \|_{L^\infty}$ intervenes in the upper bound for the operator norm of H^L and H^R on L^2 reveals that this proof actually gives

$$\|H^L\|_{\mathrm{op}}, \|H^R\|_{\mathrm{op}} \le \mathrm{const} \, (1 + \| \nabla g \|_{L^\infty})^8.$$

§3.2 THE CLIFFORD ALGEBRA VERSION OF THE $T(b)$ THEOREM

In this section we shall briefly discuss the Clifford algebra form of the $T(b)$ theorem ([MM], [DJS]) together with some of its immediate extensions and corollaries. Other versions and applications (apparently the first ones) can be found in [Se1,2]. See also [Dav2] and [GLQ].

More specifically, consider the Clifford bilinear forms

$$\langle f_1, f_2 \rangle_{b_i} := \int_{\mathbb{R}^n} f_1(x) b_i(x) f_2(x) \, dx, \qquad i = 1, 2,$$

defined for $\mathbb{C}_{(m)}$-valued functions f_1, f_2, where we assume that the functions b_1, b_2 are pseudo-accretive on $\mathbb{R}^n$ and Clifford vector-valued.

We introduce the right Clifford modules $b_i \mathcal{D}_{(m)} := \{b_i f \, ; \, f \in \mathcal{D}_{(m)}\}$, $i = 1, 2$, where, as usual, $\mathcal{D}$ stands for the class of smooth, compactly supported functions on $\mathbb{R}^n$. Similarly, we define $\mathcal{D}_{(m)} b_i$, $i = 1, 2$.

Consider T a continuous morphism of right-Clifford-modules from $b_1 \mathcal{D}_{(m)}$ into $(\mathcal{D}_{(m)} b_2)^*$ (the duality sign refers to the left-Clifford-module structure of $\mathcal{D}_{(m)} b_2$), and say that T is *associated with a standard kernel* if for some Clifford algebra valued continuous function $K(x, y)$ defined for $x \ne y$, $x, y \in \mathbb{R}^n$, and for some number δ, $0 < \delta \le 1$, one has:

(1) $|K(x,y)| \lesssim |x - y|^{-n}$, for $x \ne y$;

(2) $|K(x,y) - K(x, y_0)| + |K(y, x) - K(y_0, x)| \lesssim |y - y_0|^\delta |x - y|^{-n-\delta}$, for all $x \ne y$ with $|y - y_0| < \frac{1}{2}|y - x|$;

(3) $T(b_1 \varphi)(\psi b_2) = \iint_{\mathbb{R}^n \times \mathbb{R}^n} \psi(x) b_2(x) K(x, y) b_1(y) \varphi(y) \, dx dy$ for any $\varphi, \, \psi \in \mathcal{D}_{(m)}$ with disjoint supports.

By analogy with the previous usage, we write

$$T(b_1\varphi)(\psi b_2) = \langle \psi, T(b_1\varphi)\rangle_{b_2}.$$

T^t, the transpose of T, is the unique continuous morphism between the left Clifford modules $\mathcal{D}_{(m)}b_2$ and $(b_1\mathcal{D}_{(m)})^*$ such that, for any $\varphi, \psi \in \mathcal{D}_{(m)}$, one has

$$\langle \psi, T(b_1\varphi)\rangle_{b_2} = \langle T^t(\psi b_2), \varphi\rangle_{b_1}.$$

Note that T^t is associated with the standard kernel $K(y, x)$.

Introducing the change of variables operators $\mathcal{A}_{x,t}f := t^{-n/2}f((\cdot - x)/t)$, for $x \in \mathbb{R}^n$ and $t > 0$, we say that the T has *the weak boundedness property with respect to* b_1 *and* b_2 if the operators $\mathcal{A}_{x,t}b_2Tb_1\mathcal{A}_{x,t}{}^{-1}$ are bounded from $\mathcal{D}_{(m)}$ to $\mathcal{D}'_{(m)}$ uniformly in x and t.

If T is associated with a standard kernel, we can define $T(b_1)$ as a Clifford left-linear functional defined on the space

$$\left\{ \psi b_2 \, ; \, \psi \in \mathcal{D}_{(m)}, \, \int_{\mathbb{R}^n} \psi(x)b_2(x)\, dx = 0 \right\}$$

by setting

$$T(b_1)(\psi b_2) := \langle T(b_1\varphi), \psi)\rangle_{b_2}$$
$$+ \iint_{\mathbb{R}^n \times \mathbb{R}^n} \psi(x)b_2(x)(K(x,y) - K(x_0, y))b_1(y)(1 - \varphi(y))\, dx\, dy,$$

for some $\varphi \in \mathcal{D}$, $\varphi \equiv 1$ in a neighborhood of $\operatorname{supp}\psi$. Likewise, we define $T^t(b_2)$.

We are now almost ready to state the Clifford algebra version of the $T(b)$ theorem. To this effect, recall that a locally integrable function f belongs to $BMO(\mathbb{R}^n)$ if

$$\sup_Q \frac{1}{|Q|} \int_Q |f(x) - f_Q|\, dx < +\infty,$$

where the supremum is taken over all cubes Q in $\mathbb{R}^n$, and f_Q is the integral mean of f on Q.

54

Theorem 3.8. *With the above notations, T associated to a standard kernel has an extension as a bounded operator on $L^2(\mathbb{R}^n)_{(m)}$ if and only if*

(1) $T(b_1) \in BMO(\mathbb{R}^n)_{(m)}$;

(2) $T^t(b_2) \in BMO(\mathbb{R}^n)_{(m)}$;

(3) *T has the weak boundedness property with respect to b_1 and b_2.*

Sketch of Proof. One can readily adapt the proof of the necessity of $(1) - (3)$ from the scalar case (Peetre-Spanne-Stein; see e.g. [**Jo**]) to this somewhat more general context, therefore we consider the sufficiency. In fact, it is enough to treat only the special situation when $T(b_1) = T^t(b_2) = 0$ (in which case, the theorem is actually due to McIntosh and Meyer [**MM**]). The general case, $T(b_1)$, $T^t(b_2) \in BMO(\mathbb{R}^n)_{(m)}$ reduces to the previous one by subtracting off some paraproduct-like operators. More precisely, setting

$$\Pi_1(f) := \sum_k \triangle_k^{L,2}(T(b_1)) E_{k-1}^{L,1}(b_1^{-1} f)$$

and

$$\Pi_2(f) := \sum_k \triangle_k^{L,1}(T^t(b_2)) E_{k-1}^{L,2}(b_2^{-1} f),$$

where $\triangle_k^{L,i}$ and $E_k^{L,i}$ are defined as in section §2.5 with respect to b_i, $i = 1, 2$, we would like to conclude that the operator $S := T - \Pi_1 - \Pi_2^t$ falls within the scope of the special case we have mentioned first. It is not difficult to see that Π_1 and Π_2 are bounded operators on L^2 and that $S(b_1) = S^t(b_2) = 0$. The only problem is that this operator will not be associated with a standard kernel, but the logarithmic-rate blow up near the boundaries of dyadic cubes is good enough to control it in the same fashion as before (cf. also [**Dav2**]).

Returning to the case $T(b_1) = T^t(b_2) = 0$, let us consider the systems of Haar Clifford wavelets $\{\Phi_{j,k}^L, \Phi_{j,k}^R\}_{j,k}$ and $\{\Psi_{j,k}^L, \Psi_{j,k}^R\}_{j,k}$ associated to the pseudo-accretive functions b_1 and b_2, respectively (see §2.5). In order to estimate $\langle f, T(b_1 g)\rangle_{b_2}$, we write

$$f = \sum_{j,k} \langle f, \Phi_{j,k}^R\rangle_{b_1} \Phi_{j,k}^L,$$

$$g = \sum_{j,k} \Psi_{j,k}^R \langle \Psi_{j,k}^L, g\rangle_{b_2},$$

and then formally expand

$$\langle f, T(b_1 g) \rangle_{b_2} = \sum_{j,k} \sum_{j',k'} \langle f, \Phi_{j,k}^R \rangle_{b_1} C_{j',k'}^{j,k} \langle \Psi_{j',k'}^L, g \rangle_{b_2}, \qquad (3.10)$$

where $C_{j',k'}^{j,k} := \langle \Phi_{j,k}^L, T(b_1 \Psi_{j',k'}^R) \rangle_{b_2}$. Using the properties of these dual pairs of wavelet bases one can produce estimates for $|C_{j',k'}^{j,k}|$ analogous to the ones established in the previous section, so that Schur's lemma allows us again to conclude that the infinite matrix

$$\{ C_{j',k'}^{j,k} \}_{j',k'}^{j,k}$$

is bounded on $\ell_{\mathbb{N} \times \mathbb{N}}^2 \otimes \mathbb{C}_{(m)}$. The boundedness of T on $L^2(\mathbb{R}^n)_{(m)}$ follows then from (3.10). ∎

Remarks.

(A) Another equivalent way of expressing the conditions (1)–(3) in the above theorem, which more explicitly emphasizes the translation-dilation invariance of the theory, is

$$\int_B \left| \int_B k(x,y) b_1(y) \, dy \right| dy \lesssim |B|, \qquad (3.11)$$

$$\int_B \left| \int_B b_2(x) k(x,y) \, dx \right| dx \lesssim |B|, \qquad (3.12)$$

uniformly for all balls $B \subset \mathbb{R}^n$ (cf. also [Me]).

(B) Theorem 3.8 admits various extensions some of which we shall use in the sequel. For further reference, we note that any operator T associated with a standard kernel which is (extendible to) a bounded operator on L^2 is in fact bounded on any L^p (and even on L_ω^p, with the weight ω in the Muckenhoupt class A_p), for all $1 < p < \infty$. Also, when suitably interpreted these results continue to hold for $\mathcal{H}_{(m)}$-valued standard kernels, where $\mathcal{H}$ is some Hilbert space.

Let us illustrate these ideas by considering a generalization of the usual Cauchy kernel on a Lipschitz hyper-surface analyzed in the previous section. Although valid in more generality, we restrict ourselves to the version needed later.

Let $\Sigma := \{(g(x), x) \; ; \; x \in \mathbb{R}^n\}$, for some real-valued, Lipschitz continuous function g on $\mathbb{R}^n$. For $0 < a < \frac{\pi}{2} - \arctan(\| \nabla g \|_{L^\infty})$, take Γ_a to be the upright circular cone in the upper-half space $\mathbb{R}_+^{n+1}$ having overture a and whose vertex is at the origin of the system. Also, let $\mathcal{H}$ be a Hilbert space and $K(X,Y)$ be a $\mathcal{H}_{(n)}$-valued function defined for $Y - X \notin \Gamma_a$, satisfying

$$\|K(X,Y)\|_{(n)} \lesssim |X - Y|^{-n}, \quad \text{for } Y - X \notin \Gamma_a, \tag{3.13}$$

and having the property that for any $h \in \mathcal{H}$ there exists $\epsilon = \epsilon(h) > 0$ such that, for any $X \in \mathbb{R}^{n+1}$,

$$D\langle K(\cdot, X), h \rangle = 0 \quad \text{on } \mathbb{R}^{n+1} \setminus (-\Gamma_a - \epsilon + X), \tag{3.14}$$

$$\langle K(\cdot, X), h \rangle D = 0 \quad \text{on } \mathbb{R}^{n+1} \setminus (\Gamma_a + \epsilon + X). \tag{3.15}$$

Let us make the notation $L^p_{\mathcal{H}}(\Sigma, \omega dS)$ for the Banach space of (classes of) measurable functions on Σ which are $\mathcal{H}_{(n)}$-valued and L^p-integrable with respect to the weighted surface measure ωdS.

Theorem 3.9. *If the kernel $K(X,Y)$ of the integral operator*

$$\mathcal{T}f(X) := p.v. \int_\Sigma f(Y)K(X,Y)\, dS(Y), \quad X \in \Sigma,$$

satisfies (3.13), (3.14) and (3.15), then $\mathcal{T}$ is bounded on $L^p_{\mathcal{H}}(\Sigma, \omega dS)$ for any $1 < p < \infty$ and any $\omega \in A_p$.

In fact, as it will become more apparent in the next chapter, the continuity of the operators of the type described above can be nicely expressed in terms of some weighted Hardy spaces of monogenic functions, $\mathcal{H}^p_\omega(\Omega)$, naturally associated to Ω, the domain in $\mathbb{R}^{n+1}$ lying above Σ (see §4.1 for precise definitions). More specifically, the following holds.

Theorem 3.10. *With the above hypotheses, for any $h \in \mathcal{H}_{(n)}$, the operator*

$$Tf(X) := \int_\Sigma \langle h, f(Y)K(X,Y) \rangle\, dS(Y), \quad X \in \Omega,$$

maps $L^p_{\mathcal{H}}(\Sigma, \omega dS)$ boundedly into $\mathcal{H}^p_\omega(\Omega)$.

For kernels of the form $K(X,Y) = \Phi(X - Y)$, with $\Phi(X)$ right monogenic and satisfying $|\Phi(X)| \lesssim |X|^{-n}$ in $\mathbb{R}^{n+1} \setminus \Gamma_a$, direct proofs of these results can be found in [LMS]. We shall present here an alternative argument based on an idea of Meyer [Me] which utilizes Theorem 3.8.

Proof of Theorem 3.9. To see that $K((g(x), x), (g(y), y))$ is a standard kernel, it suffices to show that for all $h \in \mathcal{H}_{(n)}$, $\|h\|_{(n)} = 1$,

$$|\nabla_X \langle K(X, Y), h \rangle| \lesssim |X - Y|^{-n-1},$$

uniformly for $X, Y \in \Sigma$ (the estimate for $\nabla_Y K$ is completely similar). Fix $h \in \mathcal{H}_{(n)}$ with $\|h\|_{(n)} = 1$, two disjoint points X, Y in Σ, and set $d := \frac{1}{2} \operatorname{dist}(Y - X, \Gamma_a)$. Cauchy's integral representation formula (1.10) gives

$$
\begin{aligned}
|\nabla_X \langle K(X, Y), h \rangle| &\lesssim \int_{|Y - X - Z| = d} \|K(Y - Z, Y)\|_{(n)} |(\nabla E)(Y - X - Z)| \, dS(Z) \\
&\lesssim |X - Y|^{-n-1},
\end{aligned}
$$

by (3.13), since $|\nabla E| \lesssim d^{-n-1}$ and $|Z| \approx |X - Y| \approx d$ on the contour of integration.

Let us now prove that for any surface ball B, which without any loss of generality is assumed to be of the form $B_r := \Sigma \cap B((g(0), 0); r)$, one has

$$
\int_{B_r} \left\| \int_{B_r} n(Y) \langle K(X, Y), h \rangle \, dS(Y) \right\|_{(n)} dS(X) \lesssim dS(B_r), \tag{3.16}
$$

uniformly for $h \in \mathcal{H}_{(n)}$ with $\|h\|_{(n)} = 1$. To this end, using Cauchy vanishing therorem (1.12), we can deform the contour of integration in the innermost integral to $\partial C_r \setminus \Sigma$, where C_r is the cylinder $C_r := \{X + t; X \in B_r, 0 < t < r\}$. Therefore the left-hand side of (3.16) can be majorized by

$$
\int_{B_r} \int_{r + B_r} |\langle K(X, Y), h \rangle| \, dS(X) dS(Y) + \int_{B_r} \int_V |\langle K(X, Y), h \rangle| \, dS(X) d\sigma_r(Y),
$$

where V is the "vertical" part of ∂C_r and $d\sigma_r$ its canonical surface measure. Let I, II respectively denote the two terms from above.

Using $|\langle K(X, Y), h \rangle| \lesssim \|K(X, Y)\|_{(n)} \lesssim |X - Y|^{-n}$, we have that the integrand in I is $\lesssim r^{-n} \approx dS(B_r)^{-1}$, hence $I \lesssim dS(B_r)$. Also, by an appropriate change of variables,

$$
II \lesssim \int_{B_r} \int_{B_{2r} \setminus B_r} \frac{1}{|X - Y|^n} \, dx dy \lesssim r^n \left(\int_{\substack{|y| < 1 \\ y \in \mathbb{R}^n}} \int_{\substack{1 < |x| < 2 \\ x \in \mathbb{R}^n}} \frac{1}{|x - y|^n} dx dy \right),
$$

58

where the last estimate was obtained by projecting Σ onto $\mathbb{R}^n$. Since the rightmost integral is finite, the conclusion follows. ∎

Chapter 4

Hardy Spaces of Monogenic Functions

In this chapter we study the Clifford algebra version of the weighted Hardy spaces on Lipschitz domains $\mathcal{H}_\omega^p(\Omega)$, $1 < p < \infty$, $\omega \in A_p$, introduced by Kenig in [**Ke**] (cf. also [**CMM**], [**GM1-2**]). Although presented in a somewhat more implicit form, such a monogenic H^p−theory also appears in the work of Li, McIntosh and Semmes [**LMS**]. See also [**Se1**], [**Se3**].

To motivate this extension, let us recall that for $n/(n-1) < p < \infty$, say, the classical H^p space for the upper-half space (Stein-Weiss [**SW**] and Fefferman-Stein [**FS**]) is the collection of all systems of conjugate harmonic functions $u = (u_j)_{j=0}^n$ on $\mathbb{R}_+^{n+1}$ so that

$$\|u\|_{H^p} := \sup_{t>0} \left(\int_{\{t\} \times \mathbb{R}^n} |u(x)|^p \, dx \right)^{1/p} < +\infty.$$

By Proposition 1.7, this is the same as saying that the $\mathbb{R}^{n+1}$−valued, monogenic function $F = u_0 - \sum_j u_j e_j$ is uniformly L^p−integrable on hyperplanes parallel to $\partial \mathbb{R}_+^{n+1}$. Thus, it is not artificial to consider spaces of arbitrary Clifford algebra valued, monogenic functions on Lipschitz domains, subject to the same type of growth restriction as above. Actually, these monogenic Hardy spaces naturally arise in connection with several other important problems like, for instance, the boundedness, regularity and boundary behavior of Clifford algebra valued integral operators, to mention only those of concern for us in this chapter.

The layout of the chapter is as follows. Several alternative descriptions of $\mathcal{H}_\omega^p(\Omega)$, including maximal function and square function characterizations, are presented in §4.1 and §4.3. The classical Riesz theory on boundary behavior is extended to this context in §4.2. Finally, in §4.4, we indicate how one can measure the regularity of the higher dimensional Cauchy integral and related operators in terms of these Hardy spaces.

Let $g : \mathbb{R}^n \longrightarrow \mathbb{R}$ be a Lipschitz function and denote by $\Sigma \subset \mathbb{R}^{n+1}$ its graph. As usual, we assume that $\mathbb{R}^{n+1}$ is embedded in the 2^n-dimensional Clifford algebra $\mathbb{R}_{(n)}$. Set $\Omega_{\pm}$ for the domains in $\mathbb{R}^{n+1}$ which lie above and respectively below Σ. For brevity, we shall sometimes write Ω instead of Ω_+. Let n be the outward unit normal of Ω, defined dS-almost everywhere on Σ, where dS is the surface measure.

We introduce *the higher dimensional (left* and *right,* respectively) *Cauchy integral operators* by

$$\mathcal{C}^L f(X) := \frac{1}{\sigma_n} \int_\Sigma f(Y) n(Y) \frac{\overline{Y - X}}{|Y - X|^{n+1}} \, dS(Y), \quad X \in \mathbb{R}^{n+1} \setminus \Sigma,$$

$$\mathcal{C}^R f(X) := \frac{1}{\sigma_n} \int_\Sigma \frac{\overline{Y - X}}{|Y - X|^{n+1}} n(Y) f(Y) \, dS(Y), \quad X \in \mathbb{R}^{n+1} \setminus \Sigma,$$

where f is a Clifford algebra valued function on Σ.

For an arbitrary application $F : \Omega_{\pm} \longrightarrow \mathbb{C}_{(n)}$ we define its *non-tangential maximal function* $\mathcal{N}F : \Sigma \longrightarrow [0, +\infty]$ by

$$\mathcal{N}F(X) := \sup_{Y \in X \pm \Gamma_\alpha} |F(Y)|, \quad X \in \Sigma,$$

where Γ_α stands for the cone $\{(x_0, x) \in \mathbb{R}^{n+1}; x_0 > |x| \tan \alpha\}$, for a fixed $\alpha \in (0, \frac{\pi}{2})$ having $\|\nabla g\|_\infty < \tan \alpha$. We also introduce *the radial maximal functions* $F_{\mathrm{rad}} : \Sigma \longrightarrow [0, +\infty]$ as

$$F_{\mathrm{rad}}(X) := \sup_{\delta > 0} |F(X \pm \delta)|.$$

For $0 < p < \infty$ and ω a nonnegative, locally integrable function on Σ, we set

$$\|F\|_{\mathcal{H}^p_\omega} := \sup_{\delta > 0} \left\{ \int_\Sigma |F(X \pm \delta)|^p \omega(X) \, dS(X) \right\}^{1/p},$$

and define the weighted Hardy spaces of monogenic functions

$$\mathcal{H}^p_\omega(\Omega_{\pm}) := \{F \text{ left monogenic in } \Omega_{\pm}; \|F\|_{\mathcal{H}^p_\omega} < +\infty\}, \qquad (4.1)$$

$$\mathcal{K}^p_\omega(\Omega_{\pm}) := \{F \text{ right monogenic in } \Omega_{\pm}; \|F\|_{\mathcal{H}^p_\omega} < +\infty\}. \qquad (4.2)$$

Recall that $\omega \in A_p$, the Muckenhoupt class, if

$$\sup_Q \left(\frac{1}{|Q|} \int_Q \bar{\omega} \, dx \right) \left(\frac{1}{|Q|} \int_Q \bar{\omega}^{-\frac{1}{p-1}} \, dx \right)^{p-1} < +\infty,$$

where the supremum is taken over all cubes Q in $\mathbb{R}^n$ and $\bar{\omega}(x) := \omega((g(x), x))$, $x \in \mathbb{R}^n$. For the basic properties of A_p weights the reader may consult the excellent exposition in [GF].

Our first result collects several equivalent characterizations of these Hardy spaces.

Theorem 4.1. *Let $1 < p < \infty$ and $\omega \in A_p$. For a left monogenic function F in Ω, the following are equivalent:*

(1) *$F \in \mathcal{H}_\omega^p(\Omega)$;*

(2) *There exists $f \in L^p(\Sigma, \omega dS)_{(n)}$ such that F is the (right) Cauchy integral extension of f, i.e. $F(X) = C^R f(X)$ for X in Ω;*

(3) *$\mathcal{N}F \in L^p(\Sigma, \omega dS)$;*

(4) *$F_{\mathrm{rad}} \in L^p(\Sigma, \omega dS)$;*

(5) *F has a non-tangential boundary limit $F^+(X)$ at almost any point $X \in \Sigma$, i.e. there exists*

$$F^+(X) := \lim_{\substack{Y \in X + \Gamma_\alpha \\ Y \to X}} F(Y) \text{ for a.e. } X \in \Sigma,$$

and F is the (right) Cauchy integral extension of its boundary trace.

In addition, if any of the above conditions is fulfilled, we also have

$$\|F\|_{\mathcal{H}_\omega^p} \approx \|\mathcal{N}F\|_{L_\omega^p} \approx \|F^+\|_{L_\omega^p} \approx \|F_{\mathrm{rad}}\|_{L_\omega^p}. \tag{4.3}$$

Remark. Obviously, analogous results are valid for right monogenic functions, for functions defined in Ω_-, and even for $\mathcal{H}_{(n)}$-valued monogenic functions in $\Omega_\pm$, where $\mathcal{H}$ is some Hilbert space.

Proof. The nontrivial implications are $(1) \Rightarrow (2) \Rightarrow (3) \Rightarrow (5) \Rightarrow (1)$. To deal with the first one, we need the following lemma.

Lemma 4.2. *Let $F \in \mathcal{H}^p_\omega(\Omega)$ and set $F_\delta := F(\cdot + \delta)$ in $\Omega - \delta$, for $\delta > 0$. Then $F_\delta = \mathcal{C}^R F_\delta$.*

Accepting this for the moment, we may use Alaoglu's theorem for the bounded sequence $\{F_\delta\}_\delta$ in $L^p(\Sigma, \omega dS)_{(n)}$ to extract a subsequence, which we denote by the same symbol $\{F_\delta\}_\delta$, which is weakly convergent in $L^p(\Sigma, \omega dS)_{(n)}$ to a certain function $f \in L^p(\Sigma, \omega dS)_{(n)}$. Now, if we fix $X \in \Omega_+$ and let δ tend to zero in the equality

$$F_\delta(X) = \frac{1}{\sigma_n} \int_\Sigma \frac{Y - X}{|Y - X|^{n+1}} n(Y) F_\delta(Y) \, dS(Y),$$

we obtain $F(X) = \mathcal{C}^R f(X)$, as desired.

Next, we consider the implication (2) $\Rightarrow$ (3). Recall the truncated Hilbert transforms from §3.1, and the usual Hardy-Littlewood maximal operator $*$,

$$f^*(X) := \sup_{r>0} \frac{1}{dS(B_r(X) \cap \Sigma)} \int_{B_r(X) \cap \Sigma} |f(Y)| \, dS(Y), \quad X \in \Sigma.$$

It is a well-known fact that $*$ is a bounded mapping of $L^p(\Sigma, \omega dS)$ for any $1 < p < \infty$ and any $\omega \in A_p$ (cf. e.g. [Jo], [GF]).

Lemma 4.3. *With the above notations, one has*

$$\mathcal{N}(\mathcal{C}^R f)(X) \lesssim \sup_{\epsilon>0} |H^R_\epsilon f(X)| + f^*(X), \tag{4.4}$$

uniformly in $X \in \Sigma$, and $f \in L^p(\Sigma, \omega dS)_{(n)}$.

Accepting this result and recalling Cotlar's inequality

$$\sup_{\epsilon>0} |H^R_\epsilon f(X)| \lesssim (H^R f)^*(X) + f^*(X)$$

(cf. e.g. [Jo]), we see that if $F = \mathcal{C}^R f$ in Ω for some $f \in L^p(\Sigma, \omega dS)_{(n)}$, then

$$\|\mathcal{N}F\|_{L^p_\omega} = \|\mathcal{N}(\mathcal{C}^R f)\|_{L^p_\omega} \lesssim \|f\|_{L^p_\omega} < +\infty \tag{4.5}$$

by the L^p_ω-boundedness of the Hardy-Littlewood maximal operator and the Hilbert transform. Note that the above reasoning also takes care of (5) $\Rightarrow$ (1).

Finally, we prove that (1)-(4) imply (5). Using (2) and following a classical pattern (see e.g. [Jo]), we must prove the boundedness of the maximal operator associated with this type of convergence (i.e. the non-tangential convergence to the boundary) together with the almost everywhere pointwise convergence for a dense subspace of $L^p(\Sigma, \omega dS)_{(n)}$.

The first part is implicitly contained in Lemma 4.3. As for the second part, take f to be Lipschitz continuous, compactly supported on Σ, so that for any $\epsilon > 0$ we have

$$\lim_{\substack{Y \in X+\Gamma_\alpha \\ Y \to X}} C^R f(Y) = \lim_{\substack{Y \in X+\Gamma_\alpha \\ Y \to X}} \frac{1}{\sigma_n} \int_{|X-Z| \geq \epsilon} \frac{\overline{Z-Y}}{|Z-Y|^{n+1}} n(Z) f(Z)\, dS(Z)$$

$$+ \lim_{\substack{Y \in X+\Gamma_\alpha \\ Y \to X}} \frac{1}{\sigma_n} \int_{|X-Z| \leq \epsilon} \frac{\overline{Z-Y}}{|Z-Y|^{n+1}} n(Z) f(Z)\, dS(Z).$$

Since f is Lipschitz and since

$$|Y - X| \lesssim |Y - Z| \quad \text{and} \quad |Z - X| \lesssim \text{dist}\,(Z, X + \Gamma_\alpha), \tag{4.6}$$

uniformly for $X, Y \in \Sigma$, and $Z \in X + \Gamma_\alpha$, we get

$$\left| \int_{|X-Z| \leq \epsilon} \frac{\overline{Z-Y}}{|Z-Y|^{n+1}} n(Z)(f(Z) - f(Y))\, dS(Z) \right|$$

$$\approx \int_{|X-Z| \leq \epsilon} |Z - Y|^{-n} |f(Z) - f(Y)|\, dS(Z) \lesssim \epsilon \| \bigtriangledown f \|_{L^\infty},$$

hence this part goes to zero as $\epsilon \to 0$. On the other hand, by the monogenicity of the Cauchy kernel,

$$\frac{1}{\sigma_n} \int_{|X-Z| \leq \epsilon} \frac{\overline{Z-Y}}{|Z-Y|^{n+1}} n(Z)\, dS(Z) = \frac{1}{\sigma_n} \int_{\substack{|X-Z|=\epsilon \\ Z \in \Omega_-}} \frac{\overline{Z-Y}}{|Z-Y|^{n+1}} \frac{Z-X}{|Z-X|}\, d\omega_\epsilon(Z),$$

and this last integral tends to $\frac{1}{2}$ for a.e. $X \in \Sigma$ if we first make $Y \to X$ and then $\epsilon \to 0$. Summarizing, we have shown that

$$\lim_{\substack{Y \in X+\Gamma_\alpha \\ Y \to X}} C^R f(Y) = \frac{1}{2}\{f(X) + H^R f(X)\}, \quad \text{for a.e. } X \in \Sigma. \tag{4.7}$$

This proves the first part of the assertion in (5), i.e. the a.e. existence of $F^+(X)$. Since $\mathcal{N}F \in L^p(\Sigma, \omega dS)$, using Lebesgue's dominated convergence theorem one can readily see that the sequence $\{F_\delta\}_\delta$ introduced in Lemma 4.2 converges in L^p_ω to F^+. Finally, letting δ go to zero in $F_\delta = \mathcal{C}^R F_\delta$, we obtain that F can be recovered as the (right) Cauchy integral extension of its boundary trace.

Finally, to get (4.3), just use (4.5)

$$\|F^+\|_{L^p_\omega} \leq \|F\|_{\mathcal{H}^p_\omega} \leq \|F_{\mathrm{rad}}\|_{L^p_\omega} \leq \|\mathcal{N}F\|_{L^p_\omega} = \|\mathcal{N}(\mathcal{C}^R F^+)\|_{L^p_\omega} \lesssim \|F^+\|_{L^p_\omega}.$$

Hence, modulo the proofs of lemmas 4.2 and 4.3, the proof of Theorem 4.1 is complete.

∎

Proof of Lemma 4.2. Clearly, it suffices to prove the statement corresponding to $\delta = 0$, assuming that F is left monogenic in a neighborhood of $\overline{\Omega}$. To this end, let us first introduce some notation. For $r, s > 0$, let $C_{r,s}$ denote the "cylinder" $\{(t, x)\, ; \ g(x) < t < g(x) + s, \ |x| < r\}$ and set Σ_r for $\Sigma \cap C_{r,s}$. Now fix $X \in \Omega$ and choose r, s large enough such that $X \in C_{r,s}$. As F is left monogenic in a neighborhood of the closure of $C_{r,s}$, Cauchy's reproducing formula gives

$$
\begin{aligned}
F(X) = &\frac{1}{\sigma_n} \int_{\Sigma_r} \frac{\overline{Y - X}}{|Y - X|^{n+1}} n(Y) F(Y)\, dS(Y) \\
&+ \frac{1}{\sigma_n} \int_{\Sigma_r} \frac{\overline{Y - X + s}}{|Y - X + s|^{n+1}} n(Y) F(Y + s)\, dS(Y) \\
&+ \frac{1}{\sigma_n} \int_{\Sigma \cap \partial C_{r,\infty}} \int_0^s \frac{\overline{Z - X + t}}{|Z - X + t|^{n+1}} \frac{Z'}{r} F(Z + t)\, dt\, d\mu_r(Z) \\
= &: I + II + III,
\end{aligned} \tag{4.8}
$$

where Z' is the projection of $Z \in \mathbb{R}^{n+1}$ onto $\{0\} \times \mathbb{R}^n$, and μ_r is the canonical measure on $C_{r,\infty}$. Since $I \longrightarrow \mathcal{C}^R F(X)$ as $s, r \longrightarrow \infty$, all we have to check is that both II, III converge to zero when s, r tend to ∞. A simple application of Hölder's inequality gives that, if $1/p + 1/q = 1$, then

$$
\begin{aligned}
|II| &\lesssim \int_{\Sigma_r} |X - Y - s|^{-n} |F(Y + s)|\, dS(Y) \\
&\lesssim \left(\int_\Sigma |F(Y + s)|^p \omega(Y) dS(Y) \right)^{1/p} \left(\int_\Sigma |Y - X + s|^{-nq} \omega^{-q/p}(Y) dS(Y) \right)^{1/q}.
\end{aligned}
$$

The rightmost factor from above is comparable with

$$\int_{\mathbb{R}^n} \frac{\overline{\omega}^{-q/p}(y)}{(s + |x - y|)^{qn}} \, dy, \tag{4.9}$$

where x, y denote the projections of X, Y, living in $\mathbb{R}^{n+1}$, onto $\mathbb{R}^n$ identified with $\{0\} \times \mathbb{R}^n$. Estimating (4.9) is fairly standard. Since $\overline{\omega} \in A_p$, we get $\overline{\omega}^{-q/p} \in A_q$ and even $\overline{\omega}^{-q/p} \in A_{q-\epsilon}$, for some $\epsilon > 0$, and decomposing the domain of integration as a disjoint union of annuli of the form $\{y \, ; \, kt < |x - y| < (k+1)t\}$, where $k \in \mathbb{N}$, an elementary calculation reveals that this integral is $\mathcal{O}(s^{-\epsilon(\omega, p, n)})$ as $s \longrightarrow \infty$ (see also [GF], [To]). Using this and the fact that $F \in \mathcal{H}_\omega^p(\Omega)$, the conclusion is that $|II| = \mathcal{O}(s^{-\epsilon})$ for a certain small positive ϵ, uniformly in r.

The idea to estimate III is to make the "vertical" part of $\partial C_{r,s}$ "vibrate" somehow. More concretely, taking the integral average of (4.8) over the interval $[r, 2r]$, we obtain

$$F(X) = \frac{1}{r} \int_r^{2r} I \, dr + \frac{1}{r} \int_r^{2r} II \, dr + \frac{1}{r} \int_r^{2r} III \, dr.$$

The first integral still converges to $\mathcal{C}^R F(X)$, while the second one is $\mathcal{O}(s^{-\epsilon})$. To estimate the third one we need a result of a geometric nature.

Lemma 4.4. $dS \approx dr \otimes d\mu_r$ in the sense that they are absolutely continuous with respect to each other and the Radon-Nikodym derivative is (essentially) bounded by some finite, strictly positive constants from above and from bellow, respectively.

Taking this for granted for a moment, we can write

$$\left| \frac{1}{r} \int_r^{2r} III \, dr \right| \lesssim \frac{1}{r} \int_0^s \left(\int_{\Sigma_{2r} \backslash \Sigma_r} |F(Z+t)| |Z - X + t|^{-n} \, dS(Z) \right) dt$$

$$\lesssim \frac{1}{r} \int_0^s \left(\int_\Sigma |F(Z+t)|^p \omega(Z) \, dS(Z) \right)^{1/p}$$

$$\cdot \left(\int_{|x-z| \geq 1} \frac{\overline{\omega}^{-p/q}(z)}{(t + |x - z|)^{qn}} \, dz \right)^{1/q} dt.$$

Arguing as before, the product of the inner integrals is L^∞ in the variable $t \in \mathbb{R}_+$ so that, we finally get

$$\left| \frac{1}{r} \int_r^{2r} III \, dr \right| \lesssim \frac{s}{r}.$$

Letting r first, and then s, tend to ∞, we conclude the proof of the Lemma 4.2.

To prove Lemma 4.4, for any regular Borelian measure ν on Σ, let $z^*(\nu)$ be the measure defined by $z^*(\nu)(E) = \nu(z(E))$, for measurable sets $E \subseteq \mathbb{R}^n$ (recall that $z(x) := (g(x), x)$, $x \in \mathbb{R}^n$). Clearly, it suffices to check that $z^*(dr \otimes d\mu_r) \approx z^*(dS)$. First, $z^*(dS) = (1 + |\nabla g|^2)^{1/2} dx \approx dx = dr \otimes d\omega_r$ where $d\omega_r$ is the surface measure of the sphere of radius r centered at the origin of $\mathbb{R}^n$. Let $f(t)$ be a local parameterization for this sphere, so that $h(t) := (g(f(t)), f(t))$ becomes a local parameterization for $\Sigma \cap C_{r,\infty}$. It is not difficult to see that $z^*(dr \otimes d\mu_r) \approx \{\det(\langle \partial_j h, \partial_k h \rangle)_{j,k}\}^{1/2} dr \otimes dt$ and $dr \otimes d\omega_r \approx \{\det(\langle \partial_j f, \partial_k f \rangle)_{j,k}\}^{1/2} dr \otimes dt$, where $\langle \cdot, \cdot \rangle$ stands for the usual inner product in $\mathbb{R}^n$. To conclude, we need to prove that these two determinants are comparable. Set A for the $n \times (n-1)$ matrix having the vectors $\partial_j f$, $j = 1, 2, ..., n$, as columns, and B for ∇g viewed as a $1 \times n$ matrix. Since $\langle \partial_j h, \partial_k h \rangle = \langle \partial_j f, \partial_k f \rangle + \langle \nabla g, \partial_j f \rangle \langle \nabla g, \partial_k f \rangle$, the matrices of the determinants in question become AA^t and $AA^t + (AB)(AB)^t$, where the superscript t denotes the transpose. Some elementary linear algebra gives us that

$$\det A(I + BB^t)A^t = \det AA^t \det(I + BB^t),$$

and since $|\det(I + BB^t)| \approx 1$, the proof is complete. $\blacksquare$

Finally, we return to

Proof of Lemma 4.3. In the next calculation we fix an arbitrary point $X \in \Sigma$, an arbitrary point $Y \in X + \Gamma_\alpha$, and set $\epsilon := |X - Y|$. Using (4.6), we have that $|\mathcal{C}^R f(Y) - \frac{1}{2} H_\epsilon^R f(X)|$ is comparable to

$$\left| \int_\Sigma \frac{Z - Y}{|Y - Z|^{n+1}} n(Z) f(Z) \, dS(Z) - \int_{|X - Z| \geq \epsilon} \frac{Z - X}{|Z - X|^{n+1}} n(Z) f(Z) \, dS(Y) \right|$$

$$\lesssim \frac{1}{\epsilon^n} \int_{|X - Z| < \epsilon} |f(Z)| \, dS(Z) + \int_{|X - Z| \geq \epsilon} |f(Z)| \left| \frac{Z - Y}{|Z - Y|^{n+1}} - \frac{Z - X}{|Z - X|^{n+1}} \right| dS(Z).$$

The first integral above is clearly majorized by $f^*(X)$ which is of the right order. Finally, use the mean-value theorem to dominate the second integral by

$$\int_{|X - Z| \geq \epsilon} |f(Z)| \frac{\epsilon}{|X - Z|^{n+1}} \, dS(Z),$$

which in turn is further estimated as

$$\sum_{k=0}^{\infty} \int_{2^k\epsilon \le |X-Z| \le 2^{k+1}\epsilon} |f(Z)| \frac{\epsilon}{|X-Z|^{n+1}}\, dS(Z)$$

$$\lesssim \sum_{k=0}^{\infty} 2^{-k(n+1)}\epsilon^{-n} \int_{|X-Z| \le \epsilon 2^{k+1}} |f(Z)|\, dS(Z) \lesssim \sum_{k=0}^{\infty} 2^{-k} f^*(X) \approx f^*(X)$$

and the proof of the lemma is complete. ∎

Exercises.

• Using the mean value property of F, prove directly the pointwise estimate $\mathcal{N}F(X) \lesssim (F_{\mathrm{rad}})^*(X)$, uniformly for $X \in \Sigma$.

• Prove that the inclusion operator of $\mathcal{H}^p_\omega(\Omega)$ into the space of all left monogenic functions on Ω (endowed with the topology of uniform convergence on compact subsets) is compact.

• Prove the bounded domain version of Theorem 4.1.

• Part of the Theorem 4.1 also continues to hold in the case $p = 1$, $\omega \in A_1$. More precisely, for F left monogenic in Ω consider the following assertions:

(1) $\mathcal{N}F \in L^1(\Sigma, \omega dS)$;

(2) $F_{\mathrm{rad}} \in L^1(\Sigma, \omega dS)$;

(3) F has a non-tangential limit F^+ at almost any point of Σ, F^+ belongs to $L^1(\Sigma, \omega dS)_{(n)}$ and $\mathcal{C}^R F^+ = F$;

(4) There exists $f \in L^1(\Sigma, \omega dS)_{(n)}$ such that $H^R f \in L^1(\Sigma, \omega dS)_{(n)}$ and $F = \mathcal{C}^R f$ in Ω.

Prove that (1) $\Rightarrow$ (2) $\Rightarrow$ (3) $\Rightarrow$ (4). The implication (4) $\Rightarrow$ (1) appears to be still open for general Lipschitz graphs.

• Prove that for any function F which is left monogenic in Ω and has $\mathcal{N}F \in L^1(\Sigma, \omega dS)$, where $\omega \in A_1$, there holds the cancellation property $\int_\Sigma n F^+\, dS = 0$.

As the next theorem shows, these weighted Hardy spaces also turn out to be the right ranges for the higher dimensional Cauchy operators acting on $L^p(\Sigma, \omega dS)_{(n)}$.

Theorem 4.5. For $1 < p < \infty$ and ω a non-negative, locally integrable function on Σ, the higher dimensional Cauchy integral operator $\mathcal{C}^R$ maps $L^p(\Sigma, \omega dS)_{(n)}$ boundedly onto $\mathcal{H}^p_\omega(\Omega)$ if and only if ω belongs to the Muckenhoupt class A_p. An analogous statement is valid for $\mathcal{C}^L$ also.

Proof. We have already seen in the proof of Theorem 4.1 that $\mathcal{C}^R$ is a well-defined, bounded operator mapping $L^p(\Sigma, \omega dS)_{(n)}$ onto $\mathcal{H}^p_\omega(\Omega)$.

As for the converse, let us assume that, for some nonnegative locally integrable function ω on Σ, $\mathcal{C}^R : L^p(\Sigma, \omega dS)_{(n)} \longrightarrow \mathcal{H}^p_\omega(\Omega)$ is well-defined and bounded. Recall that, for any function f which is e.g. Lipschitz continuous and compactly supported on Σ, one has

$$\lim_{\delta \to +0} \mathcal{C}^R f(X + \delta) = \frac{1}{2}\{f(X) + H^R f(X)\}, \text{ for a.e. } X \in \Sigma.$$

If we now set $F_\delta(X) := \mathcal{C}^R f(X + \delta)$, $\delta > 0$, we infer that

$$\sup_{\delta > 0} \|F_\delta\|_{L^p_\omega} = \|\mathcal{C}^R f\|_{\mathcal{H}^p_\omega} \lesssim \|f\|_{L^p_\omega},$$

so that, an application of Fatou's lemma gives

$$\|(I + H^R)f\|_{L^p_\omega} \lesssim \liminf_{\delta \to 0} \|F_\delta\|_{L^p_\omega} \lesssim \|f\|_{L^p_\omega}.$$

Therefore $H^R : L^p(\Sigma, \omega dS)_{(n)} \longrightarrow L^p(\Sigma, \omega dS)_{(n)}$ is a bounded operator, hence so is $H^R(\overline{n} \cdot)$. For the remaining part of the proof we follow Coifman and Fefferman [CF]. Fix two cubes Q, Q' in $\mathbb{R}^n$ having the same side-length l and such that dist $(Q, Q') = l$, otherwise arbitrary. Also, set $Q_* := z(Q)$ and $Q'_* := z(Q')$ where $z(x) := (g(x), x)$. Now, for any $f \in L^p_{\text{comp}}(\Sigma, \omega dS)$ with supp $f \subseteq Q_*$ and any $X \in Q'_*$, we have

$$|H^R(\overline{n}f)(X)| \gtrsim \int_{Q_*} |f(Y)||X - Y|^{-n} dS(Y) \gtrsim |Q|^{-1} \int_Q |\bar{f}(x)| dx,$$

where $\bar{f}$ is the composition of f with z. Consequently,

$$\int_Q |\bar{f}|^p \overline{\omega} \, dx \gtrsim \int_{Q_*} |f|^p \omega \, dS \gtrsim \|f\|^p_{L^p_\omega} \gtrsim \|H^R(\overline{n}f)\|^p_{L^p_\omega} \gtrsim$$
$$\gtrsim \int_{Q'_*} |H^R(\overline{n}f)|^p \omega \, dS \gtrsim \overline{\omega}(Q') \left(\frac{1}{|Q|} \int_Q |\bar{f}| \, dx \right)^p. \tag{4.10}$$

Making $\bar{f} = \chi_Q$ in the above inequality we obtain $\bar{\omega}(Q) \gtrsim \bar{\omega}(Q')$, hence $\bar{\omega}(Q) \approx \bar{\omega}(Q')$, by symmetry. Plugging now $\bar{f} = \bar{\omega}^{-\frac{1}{p-1}}$ in (4.10), a direct calculation gives that

$$\left(\frac{1}{|Q|}\int_Q \bar{\omega}\,dx\right)\left(\frac{1}{|Q|}\int_Q \bar{\omega}^{-\frac{1}{p-1}}\,dx\right)^{p-1} \lesssim 1,$$

i.e. $\bar{\omega} \in A_p$. ∎

Before we conclude this section, it is important to point out that, for the upper-half space case, a substantial part of Theorem 4.1 also carries over to the range $(n-1)/n < p \le 1$. More precisely, we have the following.

Theorem 4.6. *Let $(n-1)/n < p \le 1$ and $\omega \in A_{\frac{np}{n-1}}$. Then, for a left monogenic function F in $\mathbb{R}_+^{n+1}$, the following are equivalent:*

  (1) $F \in \mathcal{H}_\omega^p(\mathbb{R}_+^{n+1})$;

  (2) $\mathcal{N}F \in L^p(\mathbb{R}^n, \omega dx)$;

  (3) $F_{\mathrm{rad}} \in L^p(\mathbb{R}^n, \omega dx)$.

In addition, if any of the above conditions is fulfiled, then F has a non-tangential limit $F^+(X)$ at almost every point $X \in \mathbb{R}^n$, and

$$\|F\|_{\mathcal{H}_\omega^p} \approx \|\mathcal{N}F\|_{L_\omega^p} \approx \|F^+\|_{L_\omega^p} \approx \|F_{\mathrm{rad}}\|_{L_\omega^p}.$$

Moreover, $\mathcal{H}_\omega^p(\mathbb{R}_+^{n+1})$ is a complete metric linear space when endowed with the distance $(F, G) \longmapsto \|F-G\|_{\mathcal{H}_\omega^p}^p$, and the embedding $\mathcal{H}_\omega^p(\mathbb{R}_+^{n+1}) \hookrightarrow C^\infty(\mathbb{R}_+^{n+1})_{(n)}$, where we have endowed the later space with the usual topology, is continuous.

As expected, a similar result is valid for right monogenic functions.

Exercise. Prove Theorem 4.6.

Hint: Recall a basic lemma due to Stein and Weiss ([**SW2**]), namely that if F is left (or right) monogenic, then $|F|^\epsilon$ is subharmonic for any $\epsilon > (n-1)/n$.

§4.2 BOUNDARY BEHAVIOR

Let $\mathcal{B}_\pm^L$ be the (non-tangential) trace operators mapping functions F from $\mathcal{H}_\omega^p(\Omega_\pm)$ into their non-tangential boundary limits $F^\pm \in L^p(\Sigma, \omega dS)_{(n)}$. Similarly, we define $\mathcal{B}_\pm^R$ acting on $\mathcal{K}_\omega^p(\Omega_\pm)$. Note that, according to Theorem 4.1, these operators are well-defined.

The next result, whose proof is implicitly contained in the previous section, describes the way these trace operators and the Cauchy operators link up.

Theorem 4.7. For $1 < p < \infty$ and $\omega \in A_p$, the following hold:

(1) $\mathcal{B}_\pm^R : \mathcal{K}_\omega^p(\Omega_\pm) \longrightarrow L^p(\Sigma, \omega dS)_{(n)}$ are bounded operators;

(2) $\mathcal{C}^L \mathcal{B}_\pm^R = I$ on $\mathcal{K}_\omega^p(\Omega_\pm)$;

(3) $\mathcal{B}_\pm^R \mathcal{C}^L = \frac{1}{2}(\pm I + H^L)$ on $L^p(\Sigma, \omega dS)_{(n)}$ (Plemelj formulae; cf. [**If**]);

(4) $\mathcal{B}_\pm^R \mathcal{C}^L = I$ on $\operatorname{Im} \mathcal{B}_\pm^R$;

(5) $\operatorname{Ker} \mathcal{C}^L = \operatorname{Im} \mathcal{B}_\mp^R$.

Similar results are valid for $\mathcal{B}_\pm^L$, too.

In the sequel, it is useful to have intrinsic characterizations of the spaces $\operatorname{Im} \mathcal{B}_\pm^L$ and $\operatorname{Im} \mathcal{B}_\pm^R$. Using (2) and (3) above, simple considerations show that in fact one has the following descriptions.

Corollary 4.8. We have

$$\operatorname{Im} \mathcal{B}_\pm^L = \{f \in L^p(\Sigma, \omega dS)_{(n)} \, ; \, H^L f = \pm f\},$$
$$\operatorname{Im} \mathcal{B}_\pm^R = \{f \in L^p(\Sigma, \omega dS)_{(n)} \, ; \, H^R f = \pm f\}.$$

It is then justified to introduce the following notations

$$\mathcal{H}_\pm^L(\Sigma, p, \omega) := \{f \in L^p(\Sigma, \omega dS)_{(n)} \, ; \, H^L f = \pm f\},$$
$$\mathcal{H}_\pm^R(\Sigma, p, \omega) := \{f \in L^p(\Sigma, \omega dS)_{(n)} \, ; \, H^R f = \pm f\}.$$

Theorem 4.7 has also some immediate consequences which deserve to be stated separately.

Corollary 4.9. With the same hypotheses as in Theorem 4.7 one has:

(1) $H^L H^L = H^R H^R = I$ on $L^p(\Sigma, \omega dS)_{(n)}$;

(2) $\langle H^L f, f' \rangle_\Sigma = -\langle f, H^R f' \rangle_\Sigma$, for $f \in L^p(\Sigma, \omega dS)_{(n)}$, $f' \in L^q(\Sigma, wdS)_{(n)}$, where $1/p + 1/q = 1$, $w := \omega^{-q/p} \in A_q$, and the pairing $\langle \cdot, \cdot \rangle_\Sigma$ defined by

$$\langle f, f' \rangle_\Sigma := \int_\Sigma f(X) n(X) f'(X) \, dS(X). \tag{4.11}$$

(3) $L^p(\Sigma, \omega dS)_{(n)} = \mathcal{H}_-^L(\Sigma, p, \omega) \oplus \mathcal{H}_+^R(\Sigma, p, \omega)$ and
$L^p(\Sigma, \omega dS)_{(n)} = \mathcal{H}_+^L(\Sigma, p, \omega) \oplus \mathcal{H}_-^R(\Sigma, p, \omega)$, (Calderón's decompositions);

(4) $\mathcal{H}_\pm^L(\Sigma, p, \omega) \simeq \mathcal{H}_\omega^p(\Omega_\pm)$, $\quad \mathcal{H}_\pm^R(\Sigma, p, \omega) \simeq \mathcal{K}_\omega^p(\Omega_\pm)$.

Next, we present a duality result.

71

Proposition 4.10. *If $1 < p, q < \infty$ are conjugate exponents and $\omega \in A_p$, $w := \omega^{-q/p} \in A_q$, then we have the isomorphisms*

$$\mathcal{H}_{\pm}^L(\Sigma, p, \omega)^* \simeq \mathcal{H}_{\mp}^R(\Sigma, q, w) \quad \text{and} \quad \mathcal{H}_{\pm}^R(\Sigma, p, \omega)^* \simeq \mathcal{H}_{\mp}^L(\Sigma, q, w).$$

Here $*$ refers to the corresponding left or right $\mathbb{C}_{(n)}$-module structure.

Proof. Starting with an arbitrary functional ϕ in $\mathcal{H}_+^L(\Sigma, p, \omega)^*$ and using the results described in the last part of Chapter 1, we see that there exists $g \in L^q(\Sigma, wdS)_{(n)}$ such that $\phi(f) = \langle f, g \rangle_\Sigma$ for all $f \in \mathcal{H}_+^L(\Sigma, p, \omega)$. Next, we use Calderón's decomposition (in Corollary 4.9) to write

$$g = g_- \oplus g_+ \in \mathcal{H}_-^R(\Sigma, q, w) \oplus \mathcal{H}_+^R(\Sigma, q, w)$$

so that, by (2) in Corollary 4.9,

$$\langle f, g_+ \rangle_\Sigma = \langle H^L f, g_+ \rangle_\Sigma = -\langle f, H^R g_+ \rangle_\Sigma = -\langle f, g_+ \rangle_\Sigma.$$

Therefore $\langle f, g_+ \rangle_\Sigma = 0$, hence $\phi(f) = \langle f, g_- \rangle_\Sigma$. Since the pairing $\langle \cdot, \cdot \rangle_\Sigma$ is non-degenerate we have that the mapping $\phi \rightsquigarrow g_-$ is well-defined and in fact this is an isomorphism between $\mathcal{H}_+^L(\Sigma, p, \omega)^*$ and $\mathcal{H}_-^R(\Sigma, q, w)$, etc. ∎

Let $\mathcal{R}^L(\Omega_\pm)$ and $\mathcal{R}^R(\Omega_\pm)$ be the left and right, respectively, Clifford modules spanned by $\{E(X - \cdot)\}_{X \in \Omega_\mp}$ (recall that $E(X) := \frac{1}{\sigma_n} \overline{X}/|X|^{n+1}$ is the usual Cauchy kernel).

Exercise. Prove the following Smirnov-type characterizations

$$\mathcal{H}_{\pm}^L(\Sigma, p, \omega) = \text{ the } L_\omega^p - \text{closure of } \{r|_\Sigma \, ; \, r \in \mathcal{R}^L(\Omega_\pm)\}$$
$$\mathcal{H}_{\pm}^R(\Sigma, p, \omega) = \text{ the } L_\omega^p - \text{closure of } \{r|_\Sigma \, ; \, r \in \mathcal{R}^R(\Omega_\pm)\}.$$

Note that, in particular, $\mathcal{R}^L(\Omega_\pm) \hookrightarrow \mathcal{H}_\omega^p(\Omega_\pm)$ densely, etc.

Exercise. Recall the usual Hardy space $H^1(\mathbb{R}^n)$. Prove that

$$\{f \in L^1(\mathbb{R}^n)_{(n)} \, ; \, H^L f \in L^1(\mathbb{R}^n)_{(n)}\} = \{f \in L^1(\mathbb{R}^n)_{(n)} \, ; \, H^R f \in L^1(\mathbb{R}^n)_{(n)}\}$$
$$= H^1(\mathbb{R}^n)_{(n)}.$$

§4.3 SQUARE FUNCTION CHARACTERIZATIONS

The main results of this section give descriptions of the Hardy spaces of monogenic functions obtained in terms of the natural extensions of the classical Lusin area-function and Littlewood-Paley g-function to the Clifford algebra framework.

Theorem 4.11. *Assume that $1 < p < \infty$ and $\omega \in A_p$. Then*

$$\|F\|_{\mathcal{H}_\omega^p} \approx \left(\int_\Sigma \left(\iint_{X \pm \Gamma_\alpha} |\partial_0 F(Y)|^2 |X - Y|^{1-n} dY \right)^{p/2} \omega(X)\, dS(X) \right)^{1/p},$$

and

$$\|F\|_{\mathcal{H}_\omega^p} \approx \left(\int_\Sigma \left(\int_0^\infty |\partial_0 F(X \pm t)|^2 t\, dt \right)^{p/2} \omega(X)\, dS(X) \right)^{1/p},$$

uniformly for $F \in \mathcal{H}_\omega^p(\Omega_\pm)$. Similar results are valid for right monogenic functions.

For an arbitrary Clifford valued function F defined in $\Omega_\pm$ we introduce *the Lusin area-function* by

$$\mathcal{A}_\pm(F)(X) := \left(\iint_{X \pm \Gamma_\alpha} |\partial_0 F(Y)|^2 |X - Y|^{1-n} dY \right)^{1/2}, \quad X \in \overline{\Omega_\pm},$$

and its radial analogue, *the Littlewood-Paley g−function*

$$g_\pm(F)(X) := \left(\int_0^\infty |\partial_0 F(X \pm t)|^2 t\, dt \right)^{1/2}, \quad X \in \overline{\Omega_\pm}.$$

Whenever clear from the context, we shall drop the subscripts $\pm$.

The key estimates nedded in the proof of the above theorem are formulated in the next lemma.

Lemma 4.12. *For any $1 < p < \infty$ and $\omega \in A_p$, one has*

$$\|f \pm H^L f\|_{L_\omega^p(\Sigma)} \lesssim \|g_\pm(\mathcal{C}^L f)\|_{L_\omega^p(\Sigma)} \lesssim \|f\|_{L_\omega^p(\Sigma)}, \tag{4.12}$$

and

$$\|f \pm H^L f\|_{L_\omega^p(\Sigma)} \lesssim \|\mathcal{A}_\pm(\mathcal{C}^L f)\|_{L_\omega^p(\Sigma)} \lesssim \|f\|_{L_\omega^p(\Sigma)}, \tag{4.13}$$

uniformly for $f \in L^p(\Sigma, \omega dS)_{(n)}$.

The proof of the above lemma is accomplished in several steps. The idea is to use the Hilbert space valued version of the Clifford T(b) from the previous chapter in a suitable context. Let us first deal with the area-function. To this end, we introduce the space $\mathcal{K}$ of (classes of) measurable functions $h : \Gamma_\alpha \to \mathbb{C}$ such that

$$\|h\|_\mathcal{K} := \left(\iint_\Gamma |h(Z)|^2 |Z|^{1-n} dZ \right)^{1/2} < +\infty,$$

and note that if we consider the operator

$$\mathcal{S}f(X)(Z) := \int_\Sigma f(Y)(\partial_0 E)(X - Y + Z) \, dS(Y), \quad X \in \Sigma, \ Z \in \Gamma_\alpha,$$

then $\mathcal{A}_+(f)(X) = \|\mathcal{S}(f \, n)(X)\|_{(n)}$, $X \in \Sigma$. Thus, the above estimate for the area function will be a consequence of the boundedness of the operator $\mathcal{S}$ from $L^p(\Sigma, \omega dS)_{(n)}$ into $L^p_\mathcal{K}(\Sigma, \omega dS)$, the space of $\mathcal{K}_{(n)}$–valued, L^p_ω–integrable functions on Σ. In turn, by Theorem 3.9, this comes down to checking that the kernel $K(X,Y)(Z) := (\partial_0 E)(X - Y + Z)$ satisfies the conditions (3.13)-(3.15). However, (3.14) and (3.15) are simple consequences of the monogenicity of the Cauchy kernel, and we are left with (3.13). By definition,

$$\|K(X,Y)\|_{(n)} = \iint_{\Gamma_\alpha} |(\partial_0 E)(X - Y + Z)|^2 |Z|^{1-n} dZ, \quad X, Y \in \Sigma,$$

and since for $Z \in \Gamma_\alpha$ we have $\max \{|Z|, |X - Y|\} \lesssim |X - Y + Z|$, we get

$$|(\partial_0 E)(X - Y + Z)|^2 \lesssim |X - Y + Z|^{-2(n+1)} \lesssim (|Z| + |X - Y|)^{-2(n+1)}.$$

Now if $Z = (t, z)$, projecting everything into $(0, \infty) \times \mathbb{R}^n$, we are led to considering the integral

$$\int_0^\infty \int_{\mathbb{R}^n} \frac{t^{1-n}}{(t^2 + |z|^2 + |X - Y|^2)^{n+1}} \, dt \, dz$$

which, integrated first with respect to z and then with respect to t, is easily seen to do not exceed (a multiple of) $|X - Y|^{-2n}$. This concludes the proof of the boundedness of $\mathcal{S}$.

The treatment of the corresponding estimate for the g-function is essentially the same. More specifically, this time we take $\mathcal{K} := L^2((0,\infty), t\, dt)$ and consider

$$\mathcal{T}_\pm^L f(X)(t) := \partial_0(\mathcal{C}^L f)(X \pm t), \quad X \in \Sigma, \ t > 0. \tag{4.14}$$

As $g(f)(X) = \|\mathcal{T}_+^L f(X)\|_{(n)}$ and since the kernel associated with the integral operator $\mathcal{T}_+^L$ is seen to satisfy (3.13), (3.14), (3.15), the conclusion is once again provided by Theorem 3.9.

To obtain the bound from below, we rely on a remarkable identity, however not surprising for the reader familiar with some elements of Littlewood-Paley theory. To state it, recall the Clifford bilinear form $\langle \cdot, \cdot \rangle_\Sigma$ introduced in (4.11). Also, let $\mathcal{T}_\pm^R$ denote the operator introduced in (4.14) in which $\mathcal{C}^R$ is used in place of $\mathcal{C}^L$.

Lemma 4.13. *Let* $1 < p, q < \infty$ *be conjugate exponents and let* $\omega \in A_p$, $w := \omega^{-q/p} \in A_q$. *Then, for any* $f \in L^p(\Sigma, \omega dS)_{(n)}$ *and any* $f' \in L^q(\Sigma, wdS)_{(n)}$, *we have*

$$\int_0^\infty \langle \mathcal{T}_+^L f, \mathcal{T}_-^R f' \rangle_\Sigma \, t \, dt = -\frac{1}{8} \langle (I + H^L) f, f' \rangle_\Sigma. \tag{4.15}$$

To see how this can be used to conclude the proof of Lemma 4.12, we write

$$|\langle (I + H^L) f, f' \rangle_\Sigma| \lesssim \int_0^\infty \int_\Sigma |\mathcal{T}_+^L f(X)(t)| |\mathcal{T}_-^R f'(X)(t)| \, t \, dt \, dS(X)$$
$$\lesssim \int_\Sigma \|\mathcal{T}_+^L f(X)\|_{(n)} \|\mathcal{T}_-^R f'(X)\|_{(n)} \, dS(X)$$
$$\lesssim \|\mathcal{T}_+^L f\|_{L_\mathcal{K}^p(\Sigma, \omega dS)} \|\mathcal{T}_-^R f'\|_{L_\mathcal{K}^q(\Sigma, wdS)}$$
$$\lesssim \|\mathcal{T}_+^L f\|_{L_\mathcal{K}^p(\Sigma, \omega dS)} \|f'\|_{L_w^q}.$$

Taking the suppremum over all $f' \in L^q(\Sigma, wdS)_{(n)}$ with $\|f'\|_{L_w^q} = 1$, we get (4.12). Finally, we note that the area-function can be handled similarly or, alternatively, one can use the point-wise estimate $g_\pm(F) \lesssim \mathcal{A}_\pm(F)$ on Σ (see e.g. [St]).

Next we return to the proof of Lemma 4.13. Here we shall adapt an one dimensional calculation from [DJS]. We proceed in a sequence of steps.

Step 1. For any $F \in \mathcal{H}_\omega^p(\Omega)$ and any fixed $t > 0$, $\partial_0 F(\cdot + t)$ belongs to $\mathcal{H}_\omega^p(\Omega)$. Moreover,

$$t(\partial_0 F)(X + t) \to 0, \text{ as } t \to \infty \text{ or } t \to 0, \text{ for a.e. } X \in \Sigma. \tag{4.16}$$

75

To justify this, observe that, by differentiating the Cauchy reproducing formula

$$F(X + t) = (C^L F^+)(X + t)$$

with respect to t and then multiplying both sides by t, we get

$$t(\partial_0 F)(X + t) = \int_\Sigma F^+(Y)n(Y)\,t(\partial_0 E)(Y - X - t)\,dS(Y)$$

(recall that $E(\cdot)$ is the Cauchy kernel). Now since $t(\partial_0 E)(Y - X - t)$ decays in Ω like the usual Poisson kernel in $\mathbb{R}^{n+1}_+$, a well-known argument gives $|t(\partial_0 F)(X + t)| \lesssim (F^+)^*(X)$, uniformly in $X \in \Sigma$ and $t > 0$. Consequently,

$$\left\| \sup_{t > 0} |t\,(\partial_0 F)(\cdot + t)| \right\|_{L^p_\omega} \lesssim \|F^+\|_{L^p_\omega}, \tag{4.17}$$

hence, in particular, $(\partial_0 F)(\cdot + t) \in \mathcal{H}^p_\omega(\Omega)$, for any $t > 0$.

The first convergence in (4.16) is easily seen by using once again the Poisson-like decay of $t(\partial_0 E)(X + t)$ in Ω. More specifically, a routine estimate gives

$$|t(\partial_0 F)(X + t)| \lesssim \|F^+\|_{L^p_\omega} \left(\int_{\mathbb{R}^n} \frac{t^q\,\varpi(y)^{-q/p}}{(|x - y|^2 + t^2)^{\frac{n+1}{2}q}}\,dy \right)^{1/q},$$

where $x \in \mathbb{R}^n$ is such that $X = (g(x), x)$. The last integral from above receives the same treatment as (4.9) so that, we finally get

$$|t(\partial_0 F)(X + t)| \lesssim t^{-\epsilon n/q}\,[\varpi^{-q/p}(B_1(x))]^{1/q}\|F^+\|_{L^p_\omega},$$

for some small, positive ϵ. This estimate yields the first part of (4.16). The limit for $t \to 0$ is a bit more subtle. First remark that, as a limiting case of the Cauchy vanishing formula (1.12),

$$\int_\Sigma n(Y)(\partial_0 E)(Y - X - t)\,dS(Y) = 0, \quad X \in \Sigma,\ t > 0.$$

Using this, one can easily check that, for all $X \in \Sigma$,

$$\lim_{t \to 0} \int_\Sigma f(Y)n(Y)\,t(\partial_0 E)(Y - X - t)\,dS(Y) = 0$$

76

if e.g. f is Lipschitz continuous, compactly supported on Σ (see Stein [St] p.62-63). Moreover, once again due to the Poisson-like behavior of $t(\partial_0 E)(X + t)$ on Ω,

$$\sup_{t>0} \left| \int_\Sigma f(Y) n(Y) t(\partial_0 E)(Y - X - t) \, dS(Y) \right| \lesssim f^*(X).$$

Since $\omega \in A_p$, we see that the maximal operator canonically associated to the type of convergence in question is bounded on $L^p(\Sigma, \omega dS)$. Thus, the usual argument completes the proof of Step 1.

Step 2. If $F \in \mathcal{H}_\omega^p(\Omega)$, then for all $X \in \Sigma$ and $t > 0$,

$$(\partial_0^2 F)(X + 2t) = - \int_\Sigma (\partial_0 F)(Y + t) \, n(Y) \, (\partial_0 E)(Y - X - t) \, dS(Y). \tag{4.18}$$

This is simply obtained by differentiating

$$(\partial_0 F)((X + s) + t) = \int_\Sigma (\partial_0 F)(Y + t) \, n(Y) \, E(Y - X - s) \, dS(Y)$$

with respect to s, and then making $s = t$.

Step 3. For any $f \in L^p(\Sigma, \omega dS)_{(n)}$ we have

$$\left\| \sup_{\epsilon, N > 0} \left| \int_\epsilon^N t \, \partial_0^2 (C^L f)(X + 2t) \, dt \right| \right\|_{L_\omega^p} \lesssim \|f\|_{L_\omega^p}, \tag{4.19}$$

and, for almost every $X \in \Sigma$,

$$\lim_{\substack{\epsilon \to +0 \\ N \to +\infty}} \int_\epsilon^N t \, \partial_0^2 (C^L f)(X + 2t) \, dt = -\frac{1}{8}(I + H^L) f(X). \tag{4.20}$$

To prove this, we integrate by parts twice

$$\int_\epsilon^N t \, \partial_0^2 (C^L f)(X + 2t) \, dt = \frac{1}{2} t \, \partial_0 (C^L f)(X + 2t) \Big|_\epsilon^N - \frac{1}{4}(C^L f)(X + 2t) \Big|_\epsilon^N.$$

Thus (4.19) is a consequence of (4.17) and (2) $\Rightarrow$ (4) in Theorem 4.11, while (4.20) follows from (4.16) and the Plemelj formulae.

Step 4. Here are the last details of the proof of Lemma 4.13. For two arbitrary functions, $f \in L^p(\Sigma, \omega dS)_{(n)}$ and $f' \in L^q(\Sigma, \omega dS)_{(n)}$, where $1/p + 1/q = 1$,

$w := \omega^{-q/p}$, let us write the identity (4.18) for $F := \mathcal{C}^L f$, multiply both sides on the right by $n(X) f'(X)$, and then integrate the resulting formula on Σ against $dS(X)$. The resulting equality reads

$$\int_\Sigma \partial_0^2 (\mathcal{C}^L f)(X + 2t)\, n(X)\, f'(X)\, dS(X)$$
$$= \int_\Sigma \partial_0 (\mathcal{C}^L f)(Y + t)\, n(Y)\, \partial_0 (\mathcal{C}^R f')(Y - t)\, dS(Y) = \langle \mathcal{T}_+^L f, \mathcal{T}_-^R f' \rangle_\Sigma.$$

All we need to do now is to integrate this identity against $\int_0^\infty t\, dt$. Then, permuting the integrals in the left-hand side and using (4.20), we immediately get (4.15) (all the technical problems have been taken care of in Step 3). ∎

Next, we we shall prove the converse of Theorem 4.11.

Theorem 4.14. *Let $1 < p < \infty$ and $\omega \in A_p$. For any left monogenic function F on Ω, the following conditions are equivalent:*

(1) $\mathcal{A}(F) \in L^p(\Sigma, \omega dS)$ *and* $\lim_{t \to \infty} F(X + t) = 0$ *for some* $X \in \Sigma$;

(2) $g(F) \in L^p(\Sigma, \omega dS)$ *and* $\lim_{t \to \infty} F(X + t) = 0$ *for some* $X \in \Sigma$;

(3) F *belongs to* $\mathcal{H}_\omega^p(\Omega)$.

Analogous results are valid for right monogenic functions as well.

Proof. We only need to show that $(1),(2) \Rightarrow (3)$. Let F be as in (2) (the reasoning for F as in (1) is completely similar). Consider the Hilbert space $\mathcal{K} := L^2((0, \infty), t\, dt)$ and the left monogenic $\mathcal{K}_{(n)}$-valued function U on Ω defined by

$$U(X)(t) := \partial_0 F(X + t), \quad X \in \Omega, \ t > 0.$$

Note that $U_{\mathrm{rad}}(X) = g(F)(X)$, hence $U_{\mathrm{rad}} \in L^p(\Sigma, \omega dS)$. According to Theorem 4.1, U has a non-tangential boundary trace on Σ, $U^+ \in L_{\mathcal{K}}^p(\Sigma, \omega dS)$, and it is easy to see that

$$U^+(Y)(t) = \partial_0 F(Y + t), \quad \text{for a.e. } Y \in \Sigma \text{ and } t > 0.$$

We now claim that

$$t\, |\partial_0 F(X + t)| \lesssim (U^+)^*(X), \tag{4.21}$$

uniformly for $t > 0$ and $X \in \Sigma$. To see this, note that there exists a constant $0 < \lambda < 1$ depending only on Ω such that $B_{\lambda t}(X + t) \subset \Omega$ for any $X \in \Sigma$ and any $t > 0$. Using

78

the mean-value theorem for monogenic functions, we have

$$|\partial_0 F(X+t)| \lesssim \frac{1}{|B_{\lambda t}(X+t)|} \iint_{B_{\lambda t}(X+t)} |\partial_0 F(W)|\, dW$$

(writing $W := Y + s$, with $Y \in \Sigma$ and $s > 0$)

$$\lesssim t^{-n-1} \int_{\Sigma \cap B_{\lambda t}(X)} \left(\int_{(1-\lambda)t}^{(1+\lambda)t} |\partial_0 F(Y+s)|\, ds \right) dS(Y)$$

(using Hölder's inequality in the innermost integral)

$$\lesssim t^{-n-1} \int_{\Sigma \cap B_{\lambda t}(X)} \left(\int_{(1-\lambda)t}^{(1+\lambda)t} |\partial_0 F(Y+s)|^2 s\, ds \right)^{1/2} dS(Y)$$

$$\lesssim t^{-n-1} \int_{\Sigma \cap B_{\lambda t}(X)} \|U^+(Y)\|_{(n)}\, dS(Y)$$

$$\lesssim t^{-1}(U^+)^*(X),$$

thus the claim. In particular, $\partial_0 F(\cdot + t) \in \mathcal{H}^p_\omega(\Omega)$ for any fixed $t > 0$. Now take $0 < \delta < N < \infty$, arbitrary otherwise. If we can prove that

$$\|F(\cdot + \delta) - F(\cdot + N)\|_{L^p_\omega} \leq \text{const} < +\infty \tag{4.22}$$

uniformly in δ, N, and that $\lim_{t \to \infty} F(\cdot + t) = 0$, then Fatou's lemma will give

$$\|F(\cdot + \delta)\|_{L^p_\omega} \lesssim \liminf_{N \to \infty} \|F(\cdot + \delta) - F(\cdot + N)\|_{L^p_\omega} \lesssim 1,$$

i.e. $F \in \mathcal{H}^p_\omega(\Omega)$ and we are done. To this end, for a fixed $X \in \Sigma$, we write

$$F(X + N) - F(X + \delta) = \int_\delta^N \partial_0 F(X + t)\, dt$$

$$= t\,\partial_0 F(X+t)\big|_\delta^N - \int_\delta^N t\,\partial_0^2 F(X+t)\, dt$$

$$=: I + II.$$

By (4.21), I above belongs to $L^p(\Sigma, \omega dS)_{(n)}$ uniformly in δ and N, so we only need to control the second term in a similar fashion. The idea is to use the fact that $\partial_0 F(\cdot + t)$

belongs to $\mathcal{H}^p_\omega(\Omega)$ and, therefore, one can still use the identity (4.18). Integrating both sides of this identity against $\int_\delta^N t\,dt$ yields

$$\int_\delta^N t\,\partial_0^2 F(X+t)\,dt = 4\int_\Sigma \int_{\delta/2}^{N/2} \partial_0 F(Y+t)\,n(Y)\,(\partial_0 E)(Y-X-t)\,t\,dt\,dS(Y)$$

and, by introducing $G(Y)(t) := \partial_0 F(Y+t)\,n(Y)\,\chi_{(\delta/2,N/2)}(t)$, for $Y \in \Sigma$, $t > 0$, and the kernel $K(X,Y)(t) := \overline{(\partial_0 E)(Y-X-t)}$, we can continue with

$$= \int_\Sigma \langle G(Y), K(X,Y)\rangle\,dS(Y)$$
$$=: \mathcal{S}G(X),$$

where the pairing $\langle \cdot, \cdot \rangle$ refers to the Hilbert space $\mathcal{K}_{(n)}$ (see §1.3). It easy to check that the integral operator $\mathcal{S}$ (or rather its formal transpose) satisfies the hypotheses of Theorem 3.10, so that

$$\left\| \int_\delta^N t\,\partial_0^2 F(X+t)\,dt \right\|_{L^p_\omega} \lesssim \|\mathcal{S}G\|_{L^p_\omega} \lesssim \|G\|_{L^p_\mathcal{K}(\Sigma,\omega dS)} \lesssim \|g(F)\|_{L^p_\omega},$$

and (4.22) follows. Furthermore, standard arguments show that $\lim_{t\to\infty} F(X+t)$ exists and is independent of $X \in \Sigma$ hence, by hypothesis, this limit is zero. The proof of Theorem 4.14 is therefore complete. ∎

We point out that quadratic estimates of the type presented in this section can in turn be used to prove L^p—boundedness results for convolution singular integral operators with even more general kernels than the Cauchy kernel and without using the T(b) theorem. See [LMS] and [LMQ].

Exercises.

• Prove an analogous "identity" to (4.15) for operators built in connection with the area-function. More precisely, for $f_1 \in L^p(\Sigma, \omega dS)_{(n)}$ and $f_2 \in L^q(\Sigma, w\,dS)_{(n)}$, consider

$$\mathcal{U}^L_\pm f_1(X)(Z) := \partial_0(\mathcal{C}^L f_1)(X \pm Z)n(W_\pm)^{1/2}n(X)^{-1/2},$$
$$\mathcal{U}^R_\pm f_2(X)(Z) := n(X)^{-1/2}n(W_\pm)^{1/2}\partial_0(\mathcal{C}^L f_2)(X \pm Z),$$

80

where $X \in \Sigma$, $Z \in \Gamma_\alpha$ and $W_\pm$ is the "projection" of $X \pm Z$ on Σ, i.e. $W_\pm$ is the unique point on Σ for which $X \pm Z - W_\pm$ is parallel to e_0. With these notations, prove that

$$\left| \iint_{\Gamma_\alpha} \langle \mathcal{U}_+^L f_1, \mathcal{U}_-^R f_2 \rangle_\Sigma \, |Z|^{1-n} dZ \right| \approx |\langle f_1, f_2 \rangle_\Sigma|. \tag{4.23}$$

Use this to give a direct proof to the area-function estimates in Lemma 4.12.

• Give a direct proof to the implication (1) $\Rightarrow$ (3) in Theorem 4.14.

• Let F be a left monogenic function in Ω, such that $g(F) \in L^p(\Sigma, \omega dS)$, where $1 < p < \infty$ and $\omega \in A_p$. Show that this implies that $\lim_{t \to \infty} F(X + t)$ exists and is independent of $X \in \Omega$.

• Prove that $\partial_0 F$ can be replaced by the whole gradient ∇F in Theorem 4.11 and Theorem 4.14.

• Show that one can also use higher order area- and $g-$ functions in Theorem 4.11 and Theorem 4.14. Formally, for a positive integer k and a function F defined on $\Omega_\pm$, the area-function of order k is given by

$$\mathcal{A}_\pm^k(F)(X) := \left(\iint_{X \pm \Gamma_\alpha} |(\partial_0^k F)(Y)|^2 |X - Y|^{2k-1-n} dY \right)^{1/2}, \quad X \in \Sigma,$$

and the $g-$function of order k is defined by

$$g_\pm^k(F)(X) := \left(\int_0^\infty |(\partial_0^k F)(X \pm t)|^2 t^{2k-1} dt \right)^{1/2}, \quad X \in \Sigma.$$

• Let P_{x_0} denote the usual Poisson kernel in $\mathbb{R}_+^{n+1}$, R_j the jth Riesz transform in $\mathbb{R}^n = \partial \mathbb{R}_+^{n+1}$,

$$R_j f(x) := \frac{2}{\sigma_n} \int_{\mathbb{R}^n} \frac{x_j - y_j}{|x - y|^{n+1}} f(y) \, dy, \quad x \in \mathbb{R}^{n+1}, \ j = 1, 2, \ldots, n.$$

Also, set $R_0 := I$ and assume that $1 < p < \infty$ and $\omega \in A_p$. Prove that a system of conjugate harmonic functions $u = (u_j)_j$ belongs to H_ω^p, the weighted version of the Hardy spaces for the upper-half space (as defined in the introduction of this chapter, but with dx replaced by ωdx) if and only if there exists $f \in L^p(\mathbb{R}^n, \omega dx)$ such that $u_j(x_0, x) = ((R_j f) * P_{x_0})(x)$ for all j. Moreover, $\|u\|_{H_\omega^p} \approx \|f\|_{L_\omega^p}$, hence H_ω^p is isomorphic with $L^p(\mathbb{R}^n, \omega dx)$.

Hint: Let $F := u_0 - \sum_j u_j e_j$ and let $f = f_0 - \sum_j f_j e_j$ be its boundary trace. Now, by Corollary 4.8, $H^L f = f$ and since $H^L = -\sum_{j=1}^n R_j e_j$, it follows that

$$H^L f = -\sum_{j=1}^n R_j(f_j) - \sum_{j=1}^n R_j(f_0) e_j + \sum_{\substack{1 \leq j,k \leq n \\ j \neq k}} R_k(f_j) e_j e_k.$$

Working component-wise, everything reduces to $f_j = R_j(f_0)$, for all j, etc.

• Let $1 < p < \infty$. For a $\mathbb{R}^{n+1}$-valued function f defined on $\mathbb{R}^n$ the following are equivalent:

(1) f is the non-tangential limit to the boundary of a function from the Stein-Weiss H^p space (regarded as a subspace of $\mathcal{H}^L(\mathbb{R}^{n+1}_+)$);

(2) $f \in L^p(\mathbb{R}^n)_{(n)}$ and $\hat{f}(x)(1 + ix/|x|) = 0$, $x \in \mathbb{R}^n$, where $\wedge$ denotes the Fourier transform, and i is the usual complex imaginary unit.

In particular, when $n = 1$ and $p = 2$, note that this contains the classical result of Paley and Wiener (see e.g. [Hf]) asserting that a function in $L^2(\mathbb{R})$ is the non-tangential boundary limit of an analytic function in the Hardy space $H^2(\mathbb{R}^2_+)$, if and only if its Fourier transform vanishes a.e. on the interval $(-\infty, 0)$.

• Prove Theorem 3.10.

• Prove Theorem 3.10 without the assumption (3.14).

Hint: Show that any K which satisfies (3.13) and (3.15) can be written as $K = \sum'_I K_I e_I$ where, for each I, K_I satisfies (3.13), (3.14) *and* (3.15). For this and related matters see [LMQ], [Ta].

§4.4 THE REGULARITY OF THE CAUCHY OPERATOR

For a Clifford algebra valued, locally integrable function f on Σ, we define $(\nabla_\Sigma f)(g(x), x) := \nabla_x[f(g(x), x)]$ in the distributional sense on $\mathbb{R}^n$. The action of ∇_Σ naturally extends to functions F defined in Ω by letting

$$(\nabla_\Sigma F)(g(x) + t, x) := \nabla_x[F(g(x) + t, x)], \quad x \in \mathbb{R}^n, \ t > 0.$$

Next, for $1 < p < \infty$ and $\omega \in A_p$, we introduce the homogeneous Sobolev space $L^{p,*}(\Sigma, \omega dS)_{(n)}$ as the vector space of all locally integrable functions f on Σ such that (each component of) $\nabla_\Sigma f$, taken in the distributional sense, belongs to $L^p(\Sigma, \omega dS)_{(n)}$. We endow this space with the "norm"

$$\|f\|_{L^{p,*}_\omega} := \left(\int_{\mathbb{R}^n} |\nabla_x[f(g(x), x)]|^p \, \bar{\omega}(x) \, dx \right)^{1/p}.$$

Also, we set $L^{p,1}(\Sigma, \omega dS)_{(n)} := L^{p,*}(\Sigma, \omega dS)_{(n)} \cap L^p(\Sigma, \omega dS)_{(n)}$ and endow this space with the obvious norm $\|f\|_{L^{p,1}_\omega} := \|f\|_{L^{p,*}_\omega} + \|f\|_{L^p_\omega}$. Clearly, $L^{p,1}(\Sigma, \omega dS)_{(n)}$ is a Banach space, whereas $L^{p,*}(\Sigma, \omega dS)_{(n)}$ is a Banach space modulo constants. Note that ∇_Σ is a well-defined operator on these spaces.

Also of interest for us are the following versions of (4.1):

$$\mathcal{H}^{p,*}_\omega(\Omega) := \{F \text{ Clifford valued, left monogenic in } \Omega\,;\ \partial_j F \in \mathcal{H}^p_\omega(\Omega) \text{ for all } j\},$$

and

$$\mathcal{H}^{p,1}_\omega(\Omega) := \mathcal{H}^{p,*}_\omega(\Omega) \cap \mathcal{H}^p_\omega(\Omega) = \{F \in \mathcal{H}^p_\omega(\Omega)\,;\ \partial_j F \in \mathcal{H}^p_\omega(\Omega), \text{ for all } j\},$$

which we endow with the natural "norms" $\|F\|_{\mathcal{H}^{p,*}_\omega} := \sum_{j=0}^n \|\partial_j F\|_{\mathcal{H}^p_\omega}$ and $\|F\|_{\mathcal{H}^{p,1}_\omega} := \|F\|_{\mathcal{H}^{p,*}_\omega} + \|F\|_{\mathcal{H}^p_\omega}$, respectively. Likewise, we define $\mathcal{K}^{p,*}_\omega(\Omega)$ and $\mathcal{K}^{p,1}_\omega(\Omega)$.

The next lemma essentially asserts that, for a function monogenic in a domain of $\mathbb{R}^{n+1}$, the derivatives in only n linearly independent directions actually control the entire gradient.

Lemma 4.15. *For any left or right monogenic function F in Ω we have that $|\nabla_\Sigma F| \approx |\nabla F|$, where ∇ stands for the usual gradient in $\mathbb{R}^{n+1}$.*

Proof. Assume that F is e.g. left monogenic. Note that $\partial_j[F(g(x) + t, x)] = \partial_0 F \partial_j g + \partial_j F$, $j = 1, ..., n$, and, thus,

$$|\nabla_\Sigma F| \gtrsim \left| \sum_{j=1}^n e_j \partial_j[F(g(x) + t, x)] \right| = \left| \partial_0 F \sum_{j=1}^n e_j \partial_j g + \sum_{j=1}^n e_j \partial_j F \right|$$

$$= |\partial_0 F| \left| 1 - \sum_{j=1}^n e_j \partial_j g \right|,$$

where the monogenicity of F has been used to derive the last equality. Now, since

$$\left| 1 - \sum_{j=1}^{n} e_j \partial_j g \right| \approx 1,$$

we obtain that $|\nabla_\Sigma F| \gtrsim |\partial_0 F|$. With this at hand, $|\partial_j F| \lesssim |\nabla_\Sigma F|$ for all j immediately follows. ∎

Lemma 4.16. *Any $F \in \mathcal{H}_\omega^{p,*}(\Omega)$ has a nontangential boundary limit $F^+(X)$ at almost any $X \in \Sigma$, the limit function belongs to $L^{p,*}(\Sigma, \omega dS)_{(n)}$ and $\nabla_\Sigma(F^+) = (\nabla_\Sigma F)^+$ in the distributional sense.*

Proof. If $F \in \mathcal{H}_\omega^{p,*}(\Omega)$, then $\mathcal{N}(\nabla F) \in L^p(\Sigma, \omega dS)$ so that $\mathcal{N}(\nabla F)(X) < +\infty$ at almost every point $X \in \Sigma$. For such a point X,

$$F(Y) = F(Z) + \int_0^1 \frac{d}{ds} F((Y - Z)s + Z) \, ds, \quad Y, Z \in \Gamma_\alpha + X.$$

Keeping Z fixed and letting Y approach X non-tangentially, Lebesgue's dominated convergence theorem ensures the convergence of $F(Y)$. Moreover, it is easy to see now that the limit function is actually locally integrable on Σ.

As for the last part in the lemma, let ψ be an arbitrary test function in $\mathbb{R}^n$. By means of Theorem 4.1 and repeated applications of the Lebesgue dominated convergence theorem, we have

$$\begin{aligned}
(\nabla_\Sigma F^+, \psi) &= -\int_{\mathbb{R}^n} \lim_{t \to 0} F(g(x) + t, x) \nabla_x \psi(x) \, dx \\
&= -\lim_{t \to 0} \int_{\mathbb{R}^n} F(g(x) + t, x) \nabla_x \psi(x) \, dx \\
&= \lim_{t \to 0} \int_{\mathbb{R}^n} \nabla_x [F(g(x) + t, x)] \psi(x) \, dx \\
&= \int_{\mathbb{R}^n} \lim_{t \to 0} \nabla_x [F(g(x) + t, x)] \psi(x) \, dx \\
&= ((\nabla_\Sigma F)^+, \psi),
\end{aligned}$$

where $(\cdot, \cdot)$ is the usual distributional pairing. ∎

Corollary 4.17. *The operator $\mathcal{B}^L$ mapping functions into their non-tangential boundary traces is well-defined and bounded from $\mathcal{H}_\omega^{p,*}(\Omega)$ into $L^{p,*}(\Sigma, \omega dS)_{(n)}$. A similar statement holds true for the action of $\mathcal{B}^L$ from $\mathcal{H}_\omega^{p,1}(\Omega)$ to $L^{p,1}(\Sigma, \omega dS)_{(n)}$.*

The main result of this section is the following.

Theorem 4.18. *Let $1 < p < \infty$ and $\omega \in A_p$. The Cauchy operators $\mathcal{C}^L, \mathcal{C}^R$ extend as bounded operators between $L^{p,*}(\Sigma, \omega dS)_{(n)}$ and $\mathcal{H}^{p,*}_\omega(\Omega)$, and between $L^{p,1}(\Sigma, \omega dS)_{(n)}$ and $\mathcal{H}^{p,1}_\omega(\Omega)$.*

Proof. Let f be a real-valued, Lipschitz continuous, compactly supported function on Σ. Also, let $\bar{f}(x) := f(g(x), x)$, $x \in \mathbb{R}^n$. In local coordinates, the right, say, Cauchy integral extension of f is given by

$$\mathcal{C}^R f(g(x) + t, x) = \frac{1}{\sigma_n} \int_{\mathbb{R}^n} \frac{(g(y) - g(x) - t, x - y)}{\{(g(y) - g(x) - t)^2 + |x - y|^2\}^{\frac{n+1}{2}}} (-1, \nabla g(y)) f(y) dy,$$

for $t > 0$ and $x \in \mathbb{R}^n$. Therefore, straightforward integrations by parts show that

$$\mathcal{C}^R f(g(x) + t, x) = \frac{1}{2} f(g(x), x)$$

$$- \sum_{j=1}^{n} \frac{1}{\sigma_n} \int_{\mathbb{R}^n} \frac{(y_j - x_j) \partial_j f(y)}{|x - y|^n} \lambda \left(\frac{g(y) - g(x) - t}{|x - y|} \right) dy$$

$$+ \sum_{j=1}^{n} \frac{e_j}{\sigma_n (n-1)} \int_{\mathbb{R}^n} \frac{\partial_j f(y)}{\{(g(y) - g(x) - t)^2 + |x - y|^2\}^{\frac{n-1}{2}}} dy$$

$$+ \sum_{1 \le j < k \le n} \frac{e_j e_k}{\sigma_n (n-1)} \int_{\mathbb{R}^n} \frac{\partial_k g(y) \partial_j f(y) - \partial_j g(y) \partial_k f(y)}{\{(g(y) - g(x) - t)^2 + |x - y|^2\}^{\frac{n-1}{2}}} dy,$$

where λ stands for the odd antiderivative of the mapping $t \mapsto (1 + t^2)^{-(n+1)/2}$.

Consequently, there exist some integral operators $\{\mathcal{R}_j\}_j$ such that

$$\nabla_\Sigma (\mathcal{C}^R f)(X) = \sum_{j=1}^{n} \mathcal{R}_j (\partial f / \partial T_j)(X), \quad X \in \Omega, \tag{4.24}$$

where $\{T_j(X)\}_j$ is an orthonormal frame for the tangent hyper-plane to Σ defined at almost any $X \in \Sigma$. An inspection of the kernels of the operators $\{\mathcal{R}_j\}_j$ shows that they can be treated via the $T(b)$ technology set up in Chapter 3 to obtain

$$\|\mathcal{N}(\mathcal{R}_j f)\|_{L^p_\omega} \lesssim \|f\|_{L^p_\omega}, \text{ for all } j.$$

Utilizing this and Lemma 4.15, we have

$$\|\mathcal{N}(\nabla \mathcal{C}^R f)\|_{L^p_\omega} \approx \|\mathcal{N}(\nabla_\Sigma \mathcal{C}^R f)\|_{L^p_\omega} \lesssim \sum_j \|\mathcal{N} \mathcal{R}_j (\partial f / \partial T_j)\|_{L^p_\omega}$$

$$\lesssim \sum_j \|(\partial f / \partial T_j)\|_{L^p_\omega} \lesssim \|\nabla_\Sigma f\|_{L^p_\omega}.$$

At this point, the usual density argument completes the proof of the theorem. $\blacksquare$

Corollary 4.19. *For $1 < p < \infty$ and $\omega \in A_p$, the Hilbert transforms H^L, H^R extend as bounded mappings of $L^{p,*}(\Sigma, \omega dS)_{(n)}$ and of $L^{p,1}(\Sigma, \omega dS)_{(n)}$.*

Proof. If f is a Lipschitz continuous, compactly supported function on Σ, then $F := C^R f$ belongs to $\mathcal{H}^{p,1}_\omega(\Omega)$ by the previous result. Also, a combination of Theorem 4.1 and Corollary 4.17 yields

$$(\nabla_\Sigma F)^+ = \frac{1}{2} \nabla_\Sigma (f + H^R f)$$

so that, once again by Theorem 4.18,

$$\|H^L f\|_{L^{p,*}_\omega} \lesssim \|\nabla_\Sigma (f + Hf)\|_{L^p_\omega} + \|\nabla_\Sigma f\|_{L^p_\omega} \lesssim \|\mathcal{N}(\nabla_\Sigma F)\|_{L^p_\omega} + \|\nabla_\Sigma f\|_{L^p_\omega} \lesssim \|f\|_{L^{p,*}_\omega}.$$

The second part follows similarly. ∎

Exercise. For $1 < p < \infty$ and $\omega \in A_p$, consider the family $\{S_\delta\}_{\delta > 0}$ of bounded operators on $L^p(\Sigma, \omega dS)_{(n)}$,

$$S_\delta f(X) := (C^L f)(X + \delta) + (C^L f)(X - \delta), \quad X \in \Sigma, \; \delta > 0.$$

Show that the family $\{S_\delta\}_{\delta > 0}$ is a one-parameter strongly continuous semi-group of operators on $L^p(\Sigma, \omega dS)_{(n)}$, and that the domain of the infinitesimal generator of the semi-group is $L^{p,1}(\Sigma, \omega dS)_{(n)}$.

Chapter 5

Applications to the Theory of Harmonic Functions

In this chapter we shall see that the techniques developed so far almost exclusively within the Clifford algebra framework have also important applications to several seemingly not directly related problems.

The departure point is §5.1, dealing with layer potential operators on Lipschitz domains, treated as close relatives of the higher dimensional Cauchy integral operator.

Some quantitative expressions of the Cauchy vanishing theorem for monogenic functions are obtained in section §5.2, before discussing the classical boundary value problems for the Laplace operator in Lipschitz domains (section §5.3). These results are due to Dahlberg, Jerison, Kenig and Verchota ([Dah1], [JK], [Ve], [DK]), and our Clifford algebra approach allows us to treat in a rather simple and unified manner the Dirichlet, Neumann and the regularity problem for the Laplacian.

Finally, in §5.1, we discuss a Clifford algebra version of the celebrated theorem of Burkholder, Gundy and Silverstein ([BGS]), essentially asserting that only "half" of a two-sided monogenic function determines the size of the entire function. As a corollary of this result, we give a simple proof of a theorem of Dahlberg [Dah2] concerning the L^p—norm equivalence of the Lusin area-function and the non-tangential maximal function of harmonic functions in Lipschitz domains, $0 < p < \infty$.

Several times in the sequel we shall specialize some of our previous results by taking $\omega \equiv 1$. Whenever the case, we shall simply omit it as an index (e.g. write $\mathcal{H}^p(\Omega)$, etc). Also, when no ambiguity is likely to produce, we shall denote the functions in $\mathcal{H}^p(\Omega)$ or $\mathcal{K}^p(\Omega)$ and their boundary traces by the same symbols.

§5.1 POTENTIALS OF SINGLE AND DOUBLE LAYERS

Consider the harmonic Hardy spaces

$$H^p_\omega(\Omega) := \{u \text{ real-valued, harmonic in } \Omega \, ; \, \mathcal{N}u \in L^p(\Sigma, \omega dS)\}$$

87

which we endow with the norm $\|u\|_{H^p_\omega} := \|\mathcal{N}u\|_{L^p_\omega}$, and

$$H^{p,*}_\omega(\Omega) := \{u \text{ real-valued, harmonic in } \Omega \,; \, \partial_j u \in H^p_\omega(\Omega), \text{ for all } j\},$$

$$H^{p,1}_\omega(\Omega) := H^p_\omega(\Omega) \cap H^{p,*}_\omega(\Omega),$$

endowed with $\|u\|_{H^{p,*}_\omega} := \sum_j \|\partial_j u\|_{H^p_\omega}$ and $\|u\|_{H^{p,1}_\omega} := \|u\|_{H^p_\omega} + \|u\|_{H^{p,*}_\omega}$, respectively.

The so called *double layer potential operator* $\mathcal{D}$ is formally defined for scalar-valued functions f on Σ by

$$\mathcal{D}f(X) := \frac{1}{\sigma_n} \int_\Sigma \frac{\langle Y - X, n(Y) \rangle}{|Y - X|^{n+1}} f(Y)\, dS(Y), \quad X \in \mathbb{R}^{n+1} \setminus \Sigma,$$

where $\langle \cdot, \cdot \rangle$ stands for the usual inner product in $\mathbb{R}^{n+1}$. Also, *the singular double layer potential operator* $\mathcal{K}$ is defined for scalar-valued functions f as the principal value singular integral

$$\mathcal{K}f(X) := \lim_{\epsilon \to 0} \frac{2}{\sigma_n} \int_{\substack{Y \in \Sigma \\ |X-Y| > \epsilon}} \frac{\langle Y - X, n(Y) \rangle}{|Y - X|^{n+1}} f(Y)\, dS(Y), \quad X \in \Sigma.$$

Theorem 5.1. *For $1 < p < \infty$ and $\omega \in A_p$, $\mathcal{K}$ is a bounded mapping of $L^p(\Sigma, \omega dS)$, $L^{p,*}(\Sigma, \omega dS)$ and $L^{p,1}(\Sigma, \omega dS)$, whereas $\mathcal{D}$ maps these spaces boundedly into $H^p_\omega(\Omega)$, $H^{p,*}_\omega(\Omega)$ and $H^{p,1}_\omega(\Omega)$, respectively.*

Furthermore, we have the classical jump-relations

$$\lim_{\substack{Y \in X \pm \Gamma_\alpha \\ Y \to X}} \mathcal{D}f(Y) = \frac{1}{2}\{\pm f(X) + \mathcal{K}f(X)\},$$

for almost any $X \in \Sigma$ and all $f \in L^p(\Sigma, \omega dS)$.

Proof. For a function $f \in L^p(\Sigma, \omega dS)$, writing out the Cauchy integral as

$$
\begin{aligned}
c^L f(X) =& \frac{1}{\sigma_n} \sum_{j=0}^n \int_\Sigma \frac{(y_j - x_j)n_j(Y)}{|X - Y|^{n+1}} f(Y)\, dS(Y) \\
&+ \sum_{j=1}^n \left\{ \frac{1}{\sigma_n} \int_\Sigma \frac{(y_j - x_j)n_0(Y) + (y_0 - x_0)n_j(Y)}{|X - Y|^{n+1}} f(Y)\, dS(Y) \right\} e_j \\
&+ \sum_{1 \le j < k \le n} \left\{ \frac{1}{\sigma_n} \int_\Sigma \frac{(y_j - x_j)n_k(Y) - (y_k - x_k)n_j(Y)}{|X - Y|^{n+1}} f(Y)\, dS(Y) \right\} e_j e_k,
\end{aligned}
$$

where $X = \sum_j x_j e_j \in \mathbb{R}^{n+1} \setminus \Sigma$, $Y = \sum_j y_j e_j \in \Sigma$ and $n = \sum_j n_j e_j \in \mathbb{R}^{n+1}$, we infer that $\mathcal{D}$ is simply the real part of $\mathcal{C}^L$ (or $\mathcal{C}^R$) when acting on scalar-valued functions.

Similarly, the singular principal value double layer potential operator on Σ is the real part of either Hilbert transform on Σ. Thus, the results in Chapter 4 yield the theorem. ∎

For $n \geq 2$, the *single layer potential operator* is defined by

$$\mathcal{S}f(X) := -\frac{1}{\sigma_n(n-1)} \int_\Sigma \frac{1}{|X-Y|^{n-1}} f(Y) \, dS(Y), \quad X \in \mathbb{R}^{n+1}. \tag{5.1}$$

Actually, as it stands, the integrand in (5.1) might not be absolutely convergent for arbitrary functions f in $L^2(\Sigma, dS)$, say. To remedy this, we shall replace the kernel $|X-Y|^{-n+1}$ by

$$\frac{1}{|X-Y|^{n-1}} - \frac{1}{|X_0-Y|^{n-1}}$$

where X_0 is an arbitrary fixed point in $\mathbb{R}^{n+1} \setminus \Sigma$. Since we shall be mainly concerned only in the gradient of $\mathcal{S}f(X)$, the particular choice of X_0 will not play any role in the sequel.

Theorem 5.2. *Let* $1 < p < \infty$ *and* $\omega \in A_p$.

(1) *The operator* $\mathcal{S}$ *is a well-defined, bounded mapping of* $L^p(\Sigma, \omega dS)$ *into* $H^{p,*}_\omega(\Omega)$ *and of* $L^p(\Sigma, \omega dS)$ *into* $L^{p,*}(\Sigma, \omega dS)$;

(2) *For any* f *in* $L^p(\Sigma, \omega dS)$,

$$\lim_{\substack{Y \in X \pm \Gamma_\alpha \\ Y \to X}} \frac{\partial \mathcal{S}f}{\partial n}(Y) := \lim_{\substack{Y \in X \pm \Gamma_\alpha \\ Y \to X}} \langle \nabla \mathcal{S}f(Y), n(X) \rangle = \frac{1}{2}\{\mp f(X) + \mathcal{K}^* f(X)\}, \tag{5.2}$$

for almost all $X \in \Sigma$, *where*

$$\mathcal{K}^* f(X) := \lim_{\epsilon \to 0} \frac{2}{\sigma_n} \int_{\substack{Y \in \Sigma \\ |X-Y| > \epsilon}} \frac{\langle X-Y, n(X) \rangle}{|X-Y|^{n+1}} f(Y) \, dS(Y), \quad X \in \Sigma,$$

is the formal transpose of $\mathcal{K}$;

(3) *For any* $f \in L^p(\Sigma, \omega dS)$ *and almost any* $X \in \Sigma$,

$$\lim_{\substack{Y \in X - \Gamma_\alpha \\ Y \to X}} \nabla_T \mathcal{S}f(Y) = \lim_{\substack{Y \in X + \Gamma_\alpha \\ Y \to X}} \nabla_T \mathcal{S}f(Y),$$

where ∇_T *is the tangential gradient operator,* $\nabla_T := \nabla - n(\partial/\partial n)$.

Proof. Differentiating under the integral sign gives

$$\overline{D}Sf(X) = \frac{1}{\sigma_n} \int_\Sigma \frac{X-Y}{|X-Y|^{n+1}} f(Y)\, dS(Y), \quad X \in \mathbb{R}^{n+1} \setminus \Sigma,$$

i.e. $\overline{D}Sf = -\mathcal{C}^L(f\bar{n}) = -\mathcal{C}^R(\bar{n}f)$, as $n\bar{n} = \bar{n}n = |n|^2 = 1$ a.e. on Σ. From this, (1) immediately follows.

As for (2), using $\operatorname{Re}\{nH^R(\bar{n}f)\} = -\mathcal{K}^*f$, we have

$$\begin{aligned}
\lim_{\substack{Y \in X \pm \Gamma_\alpha \\ Y \to X}} \frac{\partial Sf}{\partial n}(Y) &= \lim_{\substack{Y \in X \pm \Gamma_\alpha \\ Y \to X}} \langle (\overline{D}Sf)(Y), \bar{n}(X) \rangle \\
&= \lim_{\substack{Y \in X \pm \Gamma_\alpha \\ Y \to X}} \operatorname{Re}\{n(X)(\overline{D}Sf)(Y)\} \\
&= \frac{1}{2}\operatorname{Re}\{n\,[\mp \bar{n}f - H^R(\bar{n}f)]\}(X) \\
&= \frac{1}{2}\{\mp f(X) + \mathcal{K}^*f(X)\}.
\end{aligned}$$

To see (3), for $X \in \Sigma$ and $t \in \mathbb{R} \setminus \{0\}$, set $A(X+t) := n(X)\,\mathcal{C}^R(\bar{n}f)(X+t)$. A simple inspection shows that the components of $\nabla_T Sf$ coincide with those of $-(\bar{A} - \operatorname{Re}\bar{A})\bar{n}$, when restricted to the boundary. If we now recall from the Plemelj formulae (section §4.2) that the jump-discontinuities of A across the boundary occur precisely within its real part, we are done. ∎

§5.2 L^2–ESTIMATES AT THE BOUNDARY

First we note a boundary cancellation property for monogenic functions. Recall that $\mathcal{K}^2(\Omega)$ stands for the Hardy space of *right* monogenic functions in Ω with square integrable non-tangential maximal functions.

Lemma 5.3. *For any $F \in \mathcal{H}^2(\Omega)$ and any $G \in \mathcal{K}^2(\Omega)$, one has*

$$\int_\Sigma F\,n\,G\,dS = 0. \tag{5.3}$$

Proof. There are several ways to see this. One would be to use Proposition 4.10 from which our lemma directly follows. Instead, we could also use Cauchy's vanishing theorem and a limiting argument similar to the one presented in the proof of Lemma 4.2. ∎

For an arbitrary Clifford algebra valued function F we now introduce

$$F_\pm := \frac{1}{2}(F \pm \overline{F}),$$

resembling of the real part and the imaginary part, respectively, of a complex valued function. An easy corollary of Lemma 5.3 is the following.

Lemma 5.4. *For functions F in $\mathcal{H}^2(\Omega) \cap \mathcal{K}^2(\Omega)$, one has*

$$Re \int_\Sigma F(nF)_\pm \, dS = Re \int_\Sigma (Fn)_\pm F \, dS = Re \int_\Sigma F_\pm n F \, dS$$
$$= Re \int_\Sigma Fn F_\pm \, dS = \pm \frac{1}{2} \int_\Sigma Re \, n |F|^2 \, dS.$$

Proof. Everything is readily seen from $2(Fn)_\pm = Fn \pm \overline{Fn} = Fn \pm \overline{n}\,\overline{F}$, Lemma 5.3 and the identity

$$Re\,(\overline{n}\,\overline{F}\,F) = Re\,(F\overline{n}\,\overline{F}) = \frac{1}{2}Re\,(F\overline{n}\,\overline{F} + \overline{F\overline{n}\,\overline{F}}) = \frac{1}{2}Re\{F(\overline{n} + n)\overline{F}\}$$
$$= \frac{1}{2}Re\,(F\overline{F})(2\,Re\,n) = |F|^2 Re\,n.$$

∎

The main result of this section is the following.

Theorem 5.5. *We have*

$$\|F\|_{L^2(\Sigma)} \approx \|F_\pm\|_{L^2(\Sigma)} \approx \|Fn\|_{L^2(\Sigma)} \approx \|(Fn)_\pm\|_{L^2(\Sigma)} \approx \|(nF)_\pm\|_{L^2(\Sigma)}$$
$$\approx \|g(F)\|_{L^2(\Sigma)} \approx \|g(F_\pm)\|_{L^2(\Sigma)} \approx \|\mathcal{A}(F)\|_{L^2(\Sigma)} \approx \|\mathcal{A}(F_\pm)\|_{L^2(\Sigma)},$$

uniformly for $F \in \mathcal{H}^2(\Omega) \cap \mathcal{K}^2(\Omega)$.

Proof. The first four equivalences are immediate from the identities derived in the previous lemma, the fact that $Re\,n \leq C < 0$ almost everywhere on $\partial\Omega$ and Schwarz inequality.

The last three equivalences are obtained in a similar fashion, this time starting with e.g. the $\mathcal{K}_{(n)}$−valued two-sided monogenic function $U(X)(t) := \partial_0 F(X + t)$, where $\mathcal{K} := L^2((0, \infty), t dt)$ and $X \in \Omega$. Finally, Theorem 4.11 gives the missing link, and the proof is complete. ∎

It is interesting to point out that for the special case in which $F = Sf$, for some scalar-valued function $f \in L^2(\Sigma, dS)$, Theorem 5.5 reduces to the estimates used by Jerison, Kenig and Verchota which, in turn, go back to the work of Rellich [**Re**] (cf. also [**Ne**] and [**PW**]).

Our next result shows that Theorem 5.5 automatically extends for the larger range $2 - \theta_0 \le p \le 2$, for some $\theta > 0$ depending only on Ω. This is done by a purely real variable argument due to Dahlberg, Kenig and Verchota ([**DKV**], [**DK**]).

Theorem 5.6. *There exists θ_0 depending only on n and the Lipschitz character of Ω (i.e. $\| \nabla g \|_{L^\infty}$) such that, for any two-sided monogenic function F in $\mathcal{H}^p(\Omega)$ with $2 - \theta_0 \le p \le 2$, the following estimates hold*

$$\|F\|_{L^p(\Sigma)} \approx \|F_\pm\|_{L^p(\Sigma)} \approx \|(Fn)_\pm\|_{L^p(\Sigma)} \approx \|(nF)_\pm\|_{L^p(\Sigma)}.$$

Proof. We shall prove only the first equivalence, the rest being completely analogous. Fix $\epsilon > 0$, $\lambda > 0$, and set $G := F(\cdot + \epsilon)$, $\mathcal{O}_\lambda := \{X \in \Sigma \,;\, \mathcal{N}G(X) > \lambda\}$. Let us consider the "tent region"

$$T_\lambda := \bigcup_{X \in \Sigma \setminus \mathcal{O}_\lambda} (\Gamma_\alpha + X).$$

As G vanishes at infinity, $\mathcal{O}_\lambda$ is a bounded, open set so that ∂T_λ is a Lipschitz hypersurface which coincides with Σ outside of a compact subset of Σ. Also, the Lipschitz character of ∂T_λ depends only on the Lipschitz character of Σ. It is not difficult to see that $G \in \mathcal{H}^2(T_\lambda)$ (see also §5.4) so that, by Theorem 5.5,

$$\int_{\Sigma \setminus \mathcal{O}_\lambda} |G|^2 dS \le \int_{\partial T_\lambda} |G|^2 dS \lesssim \int_{\partial T_\lambda} |G_\pm|^2 dS \lesssim \int_{\Sigma \setminus \mathcal{O}_\lambda} |G_\pm|^2 dS + \int_{\partial T_\lambda \setminus \Sigma} |G_\pm|^2 dS.$$

Note that $|G_\pm| \le \mathcal{N}G \le \lambda$ on $\partial T_\lambda \setminus \Sigma$ by construction, thus, as $dS(\partial T_\lambda \setminus \Sigma) \approx dS(\mathcal{O}_\lambda)$, these estimates imply

$$\int_{\Sigma \setminus \mathcal{O}_\lambda} |G|^2 dS \lesssim \int_{\Sigma \setminus \mathcal{O}_\lambda} |G_\pm|^2 dS + \lambda^2 dS(\mathcal{O}_\lambda).$$

Therefore, if $p := 2 - \theta$, by Theorem 4.1,

$$\int_\Sigma |\mathcal{N} G|^{2-\theta} dS \lesssim \int_\Sigma |G|^{2-\theta} dS = \theta \int_0^\infty \lambda^{-\theta-1} \left(\int_{\Sigma \setminus \mathcal{O}_\lambda} |G|^2 dS \right) d\lambda$$

$$\lesssim \theta \int_0^\infty \lambda^{-\theta-1} \left(\int_{\Sigma \setminus \mathcal{O}_\lambda} |G_\pm|^2 dS \right) d\lambda + \theta \int_0^\infty \lambda^{-\theta+1} dS(\mathcal{O}_\lambda) \, d\lambda$$

$$\lesssim \int_\Sigma |G_\pm|^{2-\theta} dS + \frac{\theta}{2-\theta} \int_\Sigma |\mathcal{N} G|^{2-\theta} dS.$$

Hence, for θ sufficiently small, $\int_\Sigma |\mathcal{N} G|^{2-\theta} dS \lesssim \int_\Sigma |G_\pm|^{2-\theta} dS$ uniformly in $\epsilon > 0$. Recalling that $G = F(\cdot + \epsilon)$ and letting ϵ go to zero, we are done. ∎

The last result of this section can be regarded as the harmonic analogue of the corresponding estimates for monogenic functions from Chapter 4.

Lemma 5.7. *Let* $u := \mathcal{D} f$ *in* Ω *for some* f *in* $L^2(\Sigma, dS)$. *Then*

$$\|u\|_{L^2(\Sigma)} \approx \|\mathcal{N} u\|_{L^2(\Sigma)} \approx \|u_{\mathrm{rad}}\|_{L^2(\Sigma)} \approx \|\mathcal{A}(u)\|_{L^2(\Sigma)} \approx \|g(u)\|_{L^2(\Sigma)}. \tag{5.4}$$

Proof. Set $F := \mathcal{C}^L f \in \mathcal{H}^2(\Omega)$ and $G := \mathcal{C}^R f \in \mathcal{K}^2(\Omega)$. It is easy to see that $\overline{F} + G = 2u$ so that Lemma 5.3, Lemma 5.4 and Schwarz inequality give

$$\|F\|_{L^2(\Sigma)}^2 \lesssim \left| \int_\Sigma F \, n \, \overline{F} \, dS \right| = \left| \int_\Sigma F \, n \, (\overline{F} + G) \, dS \right| \lesssim \|F\|_{L^2(\Sigma)} \|u\|_{L^2(\Sigma)}.$$

Therefore, Theorem 4.1 yields

$$\|\mathcal{N} u\|_{L^2(\Sigma)} \lesssim \|\mathcal{N} F\|_{L^2(\Sigma)} \lesssim \|F\|_{L^2(\Sigma)} \lesssim \|u\|_{L^2(\Sigma)}.$$

Similar reasonings, this time starting with the Hilbert space valued harmonic function $U(X)(t) := \partial_0 \mathcal{D} f(X + t)$, complete the proof of the lemma. ∎

§5.3 BOUNDARY VALUE PROBLEMS FOR THE LAPLACE OPERATOR IN LIPSCHITZ DOMAINS

In this section we shall discuss the results of Dahlberg, Jerison, Kenig and Verchota ([Dah1], [JK], [DK], [Ve]) concerning the solvability of the classical boundary value problems for the Laplace operator on Lipschitz domains.

More specifically, we study the Dirichlet problem for $\triangle$, the Laplacian in $(n+1)$ coordinates, in Ω

$$(\mathbf{D}) \quad \begin{cases} \triangle u = 0 \text{ in } \Omega, \\ \mathcal{N}u \in L^2(\Sigma, dS), \\ u|_\Sigma = f \in L^2(\Sigma, dS), \end{cases}$$

the Neumann problem

$$(\mathbf{N}) \quad \begin{cases} \triangle u = 0 \text{ in } \Omega, \\ \mathcal{N}(\nabla u) \in L^2(\Sigma, dS), \\ \dfrac{\partial u}{\partial n}\bigg|_\Sigma = f \in L^2(\Sigma, dS), \end{cases}$$

and the regularity problem

$$(\mathbf{R}) \quad \begin{cases} \triangle u = 0 \text{ in } \Omega, \\ \mathcal{N}(\nabla u) \in L^2(\Sigma, dS), \\ \dfrac{\partial u}{\partial T_j}\bigg|_\Sigma = \dfrac{\partial f}{\partial T_j} \text{ for all } j, \end{cases}$$

where, in $(\mathbf{R})$, $f \in L^{2,*}(\Sigma, dS)$ and $\{T_j\}_j$ is an orthonormal frame for the tangent plane at almost every point of Σ. Also, the above boundary traces should be understood in the sense of the non-tangential limit to the boundary.

The basic idea is to look for the solutions expressed as layer potential extensions of a certain boundary density. Then everything reduces to inverting the corresponding singular integral operators.

Let us first deal with $(\mathbf{D})$. For $\phi \in L^2(\Sigma, dS)$ we set $u := \mathcal{D}\phi$ in Ω. Since $u|_\Sigma = \frac{1}{2}(I + \mathcal{K})\phi$, we must solve $\frac{1}{2}(I + \mathcal{K})\phi = f$, i.e. we have to invert the operator $I + \mathcal{K}$ on $L^2(\Sigma, dS)$.

Theorem 5.8. *The operators $I \pm \mathcal{K} : L^2(\Sigma, dS) \longrightarrow L^2(\Sigma, dS)$ are invertible.*

Proof. Actually we shall show the invertibility of $I + \mathcal{K}^*$, the formal transpose of $I + \mathcal{K}$. We first claim that

$$\|(I + \mathcal{K}^*)f\|_{L^2(\Sigma)} \approx \|(I - \mathcal{K}^*)f\|_{L^2(\Sigma)}, \tag{5.5}$$

uniformly for $f \in L^2(\Sigma, dS)$. To see this, for a fixed $f \in L^p(\Sigma, dS)$, we consider the function $F := \overline{D}Sf$ so that $F|_{\Omega_\pm} \in \mathcal{H}^2(\Omega_\pm)$. In addition, F is two-sided monogenic (in fact, even $\mathbb{R}^{n+1}$-valued). It is easy to see that $(F^\pm n)_+ = (\mp I + \mathcal{K}^*)f$ and that $(F^\pm n)_- = \nabla_T(\mathcal{S}f)$, where $F^\pm$ are the boundary traces of $F|_{\Omega_\pm}$. Consequently, Theorem 5.5 and the fact that $\nabla_T(\mathcal{S}f)$ is continuous across Σ (cf. (3) in Theorem 5.2) yield the claim.

Since

$$\|f\|_{L^2(\Sigma)} \lesssim \|(I - \mathcal{K}^*)f\|_{L^2(\Sigma)} + \|(I + \mathcal{K}^*)f\|_{L^2(\Sigma)} \lesssim \|f\|_{L^2(\Sigma)},$$

it follows that

$$\|f\|_{L^2(\Sigma)} \approx \|(I \pm \mathcal{K}^*)f\|_{L^2(\Sigma)}, \tag{5.6}$$

uniformly for f in $L^2(\Sigma, dS)$.

To conclude, we use the dilation invariance of (5.6) together with a simple form of the continuity argument. Setting T_s for the operator $I + \mathcal{K}^*$ in which the Lipschitz function $g(x)$ has been replaced by $x \mapsto sg(x)$, $0 \leq s \leq 1$, we see that T_s varies continuously with respect to s and is bounded from below uniformly in s (cf. (5.6)). Consequently, the index of T_s is independent of s. As $T_1 = I + \mathcal{K}^*$ and $T_0 = I$, the conclusion follows. ∎

Theorem 5.9. *The Dirichlet problem* **(D)** *has the unique solution*

$$u(X) := \frac{4}{\sigma_n} \int_\Sigma \frac{\langle Y - X, n(Y) \rangle}{|Y - X|^{n+1}} [(I + \mathcal{K})^{-1}f](Y) \, dS(Y), \quad X \in \Omega.$$

Proof. We are left with proving the uniqueness part. However, this will be a direct consequence of the a priori estimate

$$\|u\|_{L^2(\Sigma)} \approx \|\mathcal{N}u\|_{L^2(\Sigma)}, \tag{5.7}$$

uniformly for u in $H^2(\Omega)$. There are several ways to see (5.7). One can for instance employ Lemma 5.7 (see also the last exercise in this section). Another possibility is to use the fact that any $u \in H^2(\Omega)$ is of the form $\operatorname{Re} F$ for some $\mathbb{R}^{n+1}$-valued function from $\mathcal{H}^2(\Omega)$ (actually F is unique as $|\partial_0 F| = |\nabla (\operatorname{Re} F)|$; see also §5.4). Taking this for granted for the moment, on account of Theorem 5.5 and Theorem 4.1 we have

$$\|\mathcal{N}u\|_{L^2} \geq \|u\|_{L^2} = \|\operatorname{Re} F\|_{L^2} \approx \|F\|_{L^2} \approx \|\mathcal{N}F\|_{L^2} \geq \|\mathcal{N}u\|_{L^2}.$$

95

The proof is complete. ∎

Now we consider the Neumann problem (N). Due to the properties of the single layer potential operator described in the Theorem 5.2, it is natural to seek a solution for (N) in the form $u := \mathcal{S}\phi$, for some convenient ϕ in $L^2(\Sigma, dS)$.

Theorem 5.10. *The Neumann problem* (N) *has the unique (modulo additive constants) solution*

$$u(X) := -\frac{2}{(n-1)\sigma_n} \int_\Sigma \frac{1}{|X-Y|^{n-1}} [(-I + \mathcal{K}^*)^{-1} f](Y) \, dS(Y), \quad X \in \Omega.$$

Proof. As before, looking for solution in the form of a single layer potential, the existence part amounts to inverting the corresponding singular integral operator in $L^2(\Sigma, dS)$. In fact, this has already been done in Theorem 5.8. Furthermore, the uniqueness part will be a simple consequence of the following a priori estimate

$$\left\| \frac{\partial u}{\partial n} \right\|_{L^2(\Sigma)} \approx \|\mathcal{N}(\nabla u)\|_{L^2(\Sigma)}, \tag{5.8}$$

uniformly for $u \in H^{2,*}(\Omega)$. To see this, we set $F := \overline{D}u$, so that $DF = D\overline{D}u = \Delta u = 0$ in Ω. Thus, F is a $\mathbb{R}^{n+1}$-valued (hence, two-sided) monogenic function in Ω. Since $\mathcal{N}F = \mathcal{N}(\nabla u) \in L^2(\Sigma, dS)$, Theorem 4.1 gives that $F \in \mathcal{H}^2(\Omega) \cap \mathcal{K}^2(\Omega)$. Finally, as $\partial u/\partial n = \langle \overline{D}u, \overline{n} \rangle = \langle F, \overline{n} \rangle = (Fn)_+$, Theorem 5.5 concludes the proof of (5.8). ∎

The first step for treating the regularity problem is the following.

Theorem 5.11. *The operator* $\mathcal{S} : L^2(\Sigma, dS) \longrightarrow L^{2,*}(\Sigma, dS)$ *is invertible.*

Proof. The key element is the boundedness of $\mathcal{S}$ from below, which can be seen from

$$\|f\|_{L^2(\Sigma)} \lesssim \left\| \frac{\partial \mathcal{S}f}{\partial n} \right\|_{L^2(\Sigma)} \approx \|\nabla_T (\mathcal{S}f)\|_{L^2(\Sigma)} \approx \|\mathcal{S}f\|_{L^{2,*}(\Sigma)},$$

uniformly for $f \in L^2(\Sigma, dS)$. With this at hand, the invertibility follows from a continuity argument similar to the one used in proof of Theorem 5.8. ∎

Finally, we are in a position to prove the following.

Theorem 5.12. *The regularity problem* (R) *has the unique (modulo additive constants) solution*

$$u(X) := -\frac{1}{(n-1)\sigma_n} \int_\Sigma \frac{1}{|X-Y|^{n-1}} (\mathcal{S}^{-1}f)(Y) \, dS(Y), \quad X \in \Omega.$$

Proof. The existence is clear from Theorem 5.11, while the uniqueness follows from the a priori estimate

$$\| \nabla_T u\|_{L^2(\Sigma)} \approx \|\mathcal{N}(\nabla u)\|_{L^2(\Sigma)}, \tag{5.9}$$

uniformly in $u \in H^{2,*}(\Omega)$. As for (5.9), if we set $F := \overline{D}u$, we see that $|(\nabla_T u)| = |(Fn)_-|$, so that the conclusion is provided by Theorem 5.5 and Theorem 4.1. ∎

Exercise. Show that the oblique derivative problem

$$\begin{cases} \triangle u = 0 \text{ in } \Omega, \\ \mathcal{N}(\nabla u) \in L^2(\Sigma, dS), \\ (\partial_0 u)|_\Sigma = f \in L^2(\Sigma, dS), \end{cases}$$

has a unique solution.

Hint: Existence follows by shwoing that the operator $f \longrightarrow (\partial_0 \mathcal{S}f)|_\Sigma$ is invertible. Uniqueness is provided by the a priori estimate

$$\|\mathcal{N}(\nabla u)\|_{L^2(\Sigma)} \approx \|\partial_0 u\|_{L^2(\Sigma)}$$

which, in turn, follows from Theorem 5.5.

Exercise. Prove that the operators $\pm I + \mathcal{K}$ are invertible on $L^{2,*}(\Sigma, dS)$.

Hint: Prove the identity $\mathcal{K}\mathcal{S} = \mathcal{S}\mathcal{K}^*$.

Exercise. Show that any $u \in H^2(\Omega)$ is of the form $\mathcal{D}f$ for some scalar valued function f in $L^2(\Sigma, dS)$.

Hint: Let $\{\Omega_\nu\}_\nu$ be a nested sequence of smooth domains exhausting Ω in a suitable way. Use the maximum principle for harmonic functions to show that, with self-explanatory notation,

$$\mathcal{D}_\nu[2(I + \mathcal{K}_\nu)^{-1}(u|_{\Sigma_\nu})] = u|_{\Omega_\nu}, \quad \text{for all } \nu,$$

so that, by a weak* convergence argument, one can find $f \in L^2(\Sigma, dS)$ with $u = \mathcal{D}f$ in Ω.

Remark. In this section we have sketched the L^p theory for the boundary value problems for the Laplace operator on Ω only for $p = 2$. However, similar results are valid in L^p for certain larger ranges of p's (cf. [**Dah1**], [**DK**], [**Ve**]). In particular, the

Dirichlet problem (**D**) is uniquely solvable for any $f \in L^p(\Sigma, dS)$ with $2 - \epsilon < p < \infty$, while the same holds true for the Neumann problem in the range $1 < p < 2 + \epsilon$. Here ϵ is a small, positive constant, depending only on the domain Ω.

Note that, at least the $2 - \epsilon < p \leq 2$ part, also follows from Theorem 5.6 and the arguments above. Actually we can do better than this as Theorem 5.8 automatically extends to L^p for p in a small interval around 2. More specifically, we have the following result due to Calderón ([**Ca**]).

Theorem 5.13. *Let T be an operator which maps measurable functions on Σ into mesurable functions on Σ and is bounded on any $L^p(\Sigma, dS)$ for p near 2. If $T :$ $L^2(\Sigma, dS) \longrightarrow L^2(\Sigma, dS)$ is bounded from below, then $T : L^p(\Sigma, dS) \longrightarrow L^p(\Sigma, dS)$ is also bounded from below for p near 2.*

Note that, in particular, if T is an isomorphism of $L^2(\Sigma, dS)$, i.e. both T and T^* are bounded from below, then actually T is an isomorphism of $L^p(\Sigma, dS)$ for p in a small, open interval $(2 - \epsilon, 2 + \epsilon)$.

Proof. Let $\mathcal{L}^p$ be the Banach space of all bounded linear operators on $L^p(\Sigma, dS)$. Set $A := T^*T - \epsilon$ and $B := -\frac{1}{2}\|A\|_{\mathcal{L}^2} + A$. For some small $\epsilon > 0$, the operator A is self-adjoint and positive, hence $\|B\|_{\mathcal{L}^2} \leq \frac{1}{2}\|A\|_{\mathcal{L}^2}$. Since by the Riesz-Thorin theorem

$$\limsup_{p \to 2} \|B\|_{\mathcal{L}^p} \leq \|B\|_{\mathcal{L}^2} < \epsilon + \frac{1}{2}\|A\|_{\mathcal{L}^2},$$

it follows that $T^*T = (\epsilon + \frac{1}{2}\|A\|_{\mathcal{L}^2}) + B$ is actually invertible (via a Neumann series) in L^p for $|p - 2|$ small. From this, the conclusion easily follows. ∎

§5.4 A BURKHOLDER-GUNDY-SILVERSTEIN TYPE THEOREM
FOR MONOGENIC FUNCTIONS AND APPLICATIONS

In the classical setting of one complex variable, the theorem of Burkholder, Gundy and Silverstein ([**BGS**]) asserts that a holomorphic function belongs to the Hardy space $\mathcal{H}^p(\mathbb{R}^2_+)$ if and only if the non-tangential maximal function of its real part belongs to $L^p(\mathbb{R})$, $0 < p < \infty$.

Recall $\mathcal{A}(F)$ and $g(F)$, the are- and g-function of F, respectively (see §4.3). In this section we shall prove the following.

Theorem 5.14. *Let $0 < p < \infty$. Then, for a two-sided monogenic function F in Ω such that $\lim_{t\to\infty} F(X + t) = 0$ for some $X \in \Sigma$, the following are equivalent:*

(1) $\mathcal{N}F \in L^p(\Sigma, dS)$;

(2) $F_{\mathrm{rad}} \in L^p(\Sigma, dS)$;

(3) $\mathcal{A}(F) \in L^p(\Sigma, dS)$;

(4) $g(F) \in L^p(\Sigma, dS)$;

(5) $\mathcal{N}(F_\pm) \in L^p(\Sigma, dS)$;

(6) $(F_\pm)_{\mathrm{rad}} \in L^p(\Sigma, dS)$;

(7) $\mathcal{A}(F_\pm) \in L^p(\Sigma, dS)$;

(8) $g(F_\pm) \in L^p(\Sigma, dS)$.

In addition, if any of these conditions is fulfilled, then also

$$\|\mathcal{N}F\|_{L^p(\Sigma)} \approx \|F_{\mathrm{rad}}\|_{L^p(\Sigma)} \approx \|\mathcal{A}(F)\|_{L^p(\Sigma)} \approx \|g(F)\|_{L^p(\Sigma)}$$
$$\approx \|\mathcal{N}(F_\pm)\|_{L^p(\Sigma)} \approx \|(F_\pm)_{\mathrm{rad}}\|_{L^p(\Sigma)} \approx \|g(F_\pm)\|_{L^p(\Sigma)} \approx \|\mathcal{A}(F_\pm)\|_{L^p(\Sigma)}.$$

In particular, if $1 < p < \infty$, then F belongs to $\mathcal{H}^p(\Omega)$ and $\|F\|_{\mathcal{H}^p}$ is also equivalent with any of the above twelve L^p-norms.

We first recall some essentially well-known estimates. Recall that the superscript $*$ stands for the usual Hardy-Littlewood maximal operator.

Lemma 5.15. *For any function u harmonic in Ω one has*

$$|\nabla u(X + t)| \lesssim t^{-1}[(\mathcal{N}|u|^{p/2})^*(X)]^{2/p}, \quad 0 < p < 2,$$
$$|\nabla u(X + t)| \lesssim t^{-1}\mathcal{N}(u)^*(X),$$
$$|\nabla u(X + t)| \lesssim t^{-1-n/p}\|\mathcal{N}u\|_{L^p(\Sigma)}, \quad 0 < p < \infty,$$
$$|\nabla u(X + t)| \lesssim t^{-1}[(g(u)^{p/2})^*(X)]^{2/p}, \quad 0 < p < 2,$$
$$|\nabla u(X + t)| \lesssim t^{-1}g(u)^*(X),$$
$$|\nabla u(X + t)| \lesssim t^{-1-n/p}\|g(u)\|_{L^p(\Sigma)}, \quad 0 < p < \infty,$$

uniformly for $X \in \Sigma$ and $t > 0$.

The proof of the lemma is straightforward and goes along the same lines as in the upper-half space case presented in [**FS**].

Exercise. Prove it!

Proof of Theorem 5.14. We first treat the case $0 < p \leq 2$. The idea of proof (cf. also [Ko1, Ko2]) is to use the fact that the L^2 theory is valid in arbitrary Lipschitz domains to extend the result in the range $0 < p < 2$ via some "good λ" inequalities. Finally, for the dual range, we present an argument based on the invertibility of the double layer potential operator.

Note that for any two-sided monogenic function F, we have

$$\overline{D}F_{\pm} = \frac{1}{2}\overline{D}(F \pm \overline{F}) = \frac{1}{2}\overline{D}F = \frac{1}{2}(\overline{D} + D)F = \partial_0 F. \qquad (5.10)$$

Let us first assume that $\mathcal{N}(F_{\pm}) \in L^p(\Sigma, dS)$. A convex combination of the estimates presented in Lemma 5.15 yields that, for any $0 < \alpha < 1$,

$$|\nabla F_{\pm}(X + t)| \lesssim t^{-1-\alpha n/p}[(\mathcal{N}|F_{\pm}|^{p/2})^*(X)]^{(2-2\alpha)/p},$$

uniformly for $X \in \Sigma$ and $t > 0$. Fix $\epsilon > 0$ and set $G := F(\cdot + \epsilon)$. Since

$$G(X) = \int_0^{\infty} \overline{D}G_{\pm}(X + t)dt,$$

by the boundedness of the maximal operator

$$G_{\mathrm{rad}}(X) \leq \mathrm{const}(\epsilon, \alpha, p, n)[(\mathcal{N}|F_{\pm}|^{p/2})^*(X)]^{(2-2\alpha)/p} \in L^{p/(1-\alpha)}(\Sigma, dS).$$

As $\alpha \in (0, 1)$ is arbitrary, we can use Theorem 4.1 to infer that $G \in \mathcal{H}^2(\Omega)$.

From now on, we shall keep the notations from the proof of the Theorem 5.6 with only one exception, namely $\mathcal{O}_\lambda$, which is taken to be this time

$$\mathcal{O}_\lambda := \{X \in \Sigma \,;\, \mathcal{N}(G_{\pm})(X) > \lambda\}.$$

By the above reasoning, $G \in \mathcal{H}^2(T_\lambda)$ so that, by Theorem 4.1 and Theorem 5.5, in which we take $\mathcal{N}'$ to be the corresponding nontangential maximal operator for the Lipschitz domain T_λ (i.e. $\mathcal{N}'$ corresponds to a "sharper" cone $\Gamma_{\alpha'}$), we have

$$\int_{\Sigma \backslash \mathcal{O}_\lambda} |\mathcal{N}'G|^2 dS \leq \int_{\partial T_\lambda} |G|^2 dS \lesssim \int_{\partial T_\lambda} |G_{\pm}|^2 dS \lesssim \int_{\Sigma \backslash \mathcal{O}_\lambda} |G_{\pm}|^2 dS + \int_{\partial T_\lambda \backslash \Sigma} |G_{\pm}|^2 dS.$$

Note that, once again by construction, $|G_\pm| \leq \mathcal{N}(G_\pm) \leq \lambda$ on $\partial T_\lambda \setminus \Sigma$, and that

$$\int_{\Sigma \setminus \mathcal{O}_\lambda} |G_\pm|^2 dS \leq \int_{\Sigma \setminus \mathcal{O}_\lambda} |\mathcal{N}(G_\pm)|^2 dS \leq 2 \int_0^\lambda t \, dS(\mathcal{O}_t) \, dt.$$

Thus, as before, the above estimates amount to

$$\int_{\Sigma \setminus \mathcal{O}_\lambda} |\mathcal{N}'G|^2 dS \lesssim \int_0^\lambda t \, dS(\mathcal{O}_t) \, dt + \lambda^2 dS(\mathcal{O}_\lambda).$$

Finally, we use Chebyshev's inequality to transform this into the weak type estimate

$$dS(\{X \in \Sigma; \, |\mathcal{N}'G(X)| > \lambda\}) \lesssim dS(\mathcal{O}_\lambda) + \lambda^{-2} \int_0^\lambda t \, dS(\mathcal{O}_t) \, dt$$

which, in turn, after multiplication with λ^{p-1} and integration against $\int_0^\infty d\lambda$, yields

$$\|\mathcal{N}F(\cdot + \epsilon)\|_{L^p(\Sigma)} \lesssim \|\mathcal{N}'F(\cdot + \epsilon)\|_{L^p(\Sigma)} \lesssim \|\mathcal{N}F_\pm(\cdot + \epsilon)\|_{L^p(\Sigma)}.$$

Letting $\epsilon \longrightarrow 0$ and using Lebesgue's monotone convergence theorem, we obtain the equivalence $\|\mathcal{N}F\|_{L^p(\Sigma)} \approx \|\mathcal{N}(F_\pm)\|_{L^p(\Sigma)}$.

Next, we turn our attention to the area- and $g-$ function. First, with $\mathcal{A}(G)$ instead of G and $\mathcal{N}G$ in place of $\mathcal{N}(G_\pm)$, the same arguments as before give that

$$\|g(F)\|_{L^p(\Sigma)} \lesssim \|\mathcal{A}(F)\|_{L^p(\Sigma)} \lesssim \|\mathcal{N}F\|_{L^p(\Sigma)}.$$

For the converse inequalities we follow the same line, taking this time $g(G)$ to stand for $G_\pm$, so that, we arrive at $\|\mathcal{N}G\|_{L^p(\Sigma)} \lesssim \|\mathcal{N}g(G)\|_{L^p(\Sigma)}$. Therefore, by Lebesgue's monotone convergence theorem it suffices to show that

$$\|\mathcal{N}g(G)\|_{L^p(\Sigma)} \lesssim \|g(G)\|_{L^p(\Sigma)}.$$

To this effect, we once again consider the Hilbert space $\mathcal{H} := L^2((0, +\infty), t\,dt)$, and the $\mathcal{H}_{(n)}-$valued harmonic function U, defined in Ω by $U(X)(t) := \partial_0 G(X + t)$. It is not difficult to see that

$$[\mathcal{N}U]^{p/2}(X) \lesssim [(U_{\mathrm{rad}})^{p/2}]^*(X), \quad X \in \Sigma$$

101

(cf. e.g. [FS] p.170). Since $\|U(X)\|_{(n)} = g(G)(X)$, we infer $U_{\mathrm{rad}}(X) \lesssim g(G)(X)$. Consequently,

$$\|\mathcal{N}g(G)\|_{L^p(\Sigma)} \lesssim \|U_{\mathrm{rad}}\|_{L^p(\Sigma)} \lesssim \|g(G)\|_{L^p(\Sigma)},$$

by the L^2–boundedness of the maximal operator.

At this point, we have shown that (5) implies (1)-(8). If we now assume that (8) holds true, then (5.10) and standard arguments show that $g(F) \in L^p(\Sigma, dS)$. From this point on we proceed as before, with no essential alteration, and obtain that (1)-(8) are true. The other implications, as well as the various L^p–norm equivalences, are either simple or are immediately implied by what we have proved so far, and this completes the proof of the theorem in this case.

Next, we treat the case $2 < p < \infty$. In this situation, everything is readily seen from Theorem 4.1 and Theorem 4.14 except that (5) implies any of (1)-(4) or (6)-(8). Suppose that $\mathcal{N}F_\pm \in L^p(\Sigma, dS)$. From the uniqueness in the Dirichlet problem with L^p–datum and a standard weak* compactness argument we infer that $F_\pm = \mathcal{D}f^\pm$ in Ω, for some scalar valued $f^\pm \in L^p(\Sigma, dS)$. Using this, we get

$$\|f^\pm\|_{L^p(\Sigma)} \lesssim \|(I + \mathcal{K})f^\pm\|_{L^p(\Sigma)} \approx \|F_\pm\|_{L^p(\Sigma)} \lesssim \|\mathcal{N}F_\pm\|_{L^p(\Sigma)}.$$

Therefore, as $\mathcal{D} = \operatorname{Re}\mathcal{C}^L$, we may write

$$\|\mathcal{A}(F_\pm)\|_{L^p(\Sigma)} \lesssim \|\mathcal{A}(\mathcal{C}^L f^\pm)\|_{L^p(\Sigma)} \lesssim \|\mathcal{N}(\mathcal{C}^L f^\pm)\|_{L^p(\Sigma)}$$

$$\lesssim \|f^\pm\|_{L^p(\Sigma)} \lesssim \|\mathcal{N}(F_\pm)\|_{L^p(\Sigma)}.$$

Consequently, (5.10) and standard arguments imply that

$$\|\mathcal{A}(F)\|_{L^p(\Sigma)} \approx \|\mathcal{A}(F_\pm)\|_{L^p(\Sigma)} \lesssim \|\mathcal{N}(F_\pm)\|_{L^p(\Sigma)} < +\infty.$$

Thus, by Theorem 4.14, $F \in \mathcal{H}^p(\Omega)$ and $\|\mathcal{A}(F)\|_{L^p} \lesssim \|\mathcal{N}(F_\pm)\|_{L^p} \lesssim \|\mathcal{N}F\|_{L^p} \lesssim \|\mathcal{A}F\|_{L^p}$, which completes the proof of the equivalences in the $2 < p < \infty$ case. Note that because $I + \mathcal{K}$ is actually invertible on $L^p(\Sigma, dS)$ for $2 - \epsilon < p < \infty$, the constants appearing in the above equivalences do not blow up as p approaches 2.

Finally, for the last part of the theorem, we once again invoke Theorem 4.1 and we are done. ∎

An useful observation is that a $\mathbb{R}^{n+1}$—valued monogenic function is automatically two-sided monogenic (see Proposition 1.7) so that the above theorem is valid for such functions.

As an application, we shall give a simple proof of the graph version of a theorem of Dahlberg [**Dah2**] concerning the norm equivalence between the area- and the nontangential maximal function of a harmonic function in a Lipschitz domain.

First we need the following.

Lemma 5.16. *Let $0 < p < \infty$ and let u be a harmonic function in Ω such that $\mathcal{N}u \in L^p(\Sigma, dS)$. Then there exists a unique $\mathbb{R}^{n+1}$—valued monogenic function F in Ω which dies at infinity and such that $\operatorname{Re} F = u$. The same conclusion is valid if $u_{\mathrm{rad}} \in L^p(\Sigma, dS)$, and even if $\lim_{t\to\infty} u(X + t) = 0$ and $\mathcal{A}(u) \in L^p(\Sigma, dS)$, or $\lim_{t\to\infty} u(X + t) = 0$ and $g(u) \in L^p(\Sigma, dS)$, respectively.*

Proof. By Proposition 1.7, we can construct F in Ω by setting

$$F(X + t) := -\int_t^\infty (\overline{D}u)(X + s)\, ds,$$

where $X \in \Sigma$ and $t > 0$. ∎

An easy consequence of this and Theorem 4.1, which is worth mentioning, is the well-known fact that any harmonic function in $H^p(\Omega)$, $1 < p < \infty$, has a nontangential boundary trace on Σ.

The important thing, see from Theorem 5.14, is that F has roughly the same "size" as u, i.e. $\|\mathcal{N}F\|_{L^p} \approx \|\mathcal{N}u\|_{L^p}$ (or $\|F_{\mathrm{rad}}\|_{L^p} \approx \|u_{\mathrm{rad}}\|_{L^p}$, $\|\mathcal{A}(F)\|_{L^p} \approx \|\mathcal{A}(u)\|_{L^p}$, $\|g(F)\|_{L^p} \approx \|g(u)\|_{L^p}$, respectively).

An immediate corollary of Theorem 5.14 and Lemma 5.16 is the following version of the result of Dahlberg [**Dah2**] alluded before.

Corollary 5.17. *Let u be harmonic in Ω and normalized such that $\lim_{t\to\infty} u(X+t) = 0$ for some $X \in \Sigma$. Then, for $0 < p < \infty$, the following are equivalent:*

(1) $\mathcal{N}u \in L^p(\Sigma, dS)$;

(2) $u_{\mathrm{rad}} \in L^p(\Sigma, dS)$;

(3) $\mathcal{A}(u) \in L^p(\Sigma, dS)$;

(4) $g(u) \in L^p(\Sigma, dS)$.

In addition, if any of these conditions is satisfied, then

$$\|\mathcal{N}u\|_{L^p(\Sigma)} \approx \|u_{\text{rad}}\|_{L^p(\Sigma)} \approx \|\mathcal{A}(u)\|_{L^p(\Sigma)} \approx \|g(u)\|_{L^p(\Sigma)}.$$

Finally, we discuss one more application, also due to Dahlberg [**Dah2**] (actually it was Kenig who first realized that Dahlberg's square function estimates for harmonic functions can be obtained from the much simpler square function estimates for monogenic functions [**Mc1**]).

Corollary 5.18. *For any harmonic function u in Ω which dies at infinity, one has*

$$\int_{\Sigma} |u|^2 \, dS \approx \iint_{\Omega} | \bigtriangledown u|^2 \text{dist}\, (X, \Sigma) \, d\text{Vol}.$$

Proof. Obviously, this is an equivalent formulation of $\|u\|_{L^2(\Sigma)} \approx \|g(u)\|_{L^2(\Sigma)}$, which is proved e.g. in Theorem 5.5. ∎

Exercise. For $1 < p < \infty$ and $\omega \in A_p$, prove a weighted version of Lemma 5.16 and Corollary 5.17 (it should be pointed out that these results are actually valid for $0 < p < \infty$ and $\omega \in A_\infty$; cf. [**Dah2**], [**BM**]).

Remark. For $1 < p < \infty$ and $\omega \in A_p$, a proof of the fact that $\mathcal{N}u \in L^p(\mathbb{R}^n, \omega dx)$ implies $\mathcal{A}(u) \in L^p(\mathbb{R}^n, \omega dx)$ for the upper-half space case can be seen more directly from the fact that any $u \in H^p_\omega(\mathbb{R}^{n+1}_+)$ is of the form $u = P_t * f$, $f \in L^p(\mathbb{R}^n, \omega dx)$ (with P_t standing for the usual Poisson kernel; see e.g. [**GW**] p.117). Indeed, setting $F := P_t * (f + H^L f)$ we have that F is left monogenic in $\mathbb{R}^{n+1}_+$, vanishes at infinity and $|\partial_0 F| = | \bigtriangledown u|$. Thus, by Theorem 4.14,

$$\|\mathcal{A}(u)\|_{L^p_\omega} = \|\mathcal{A}(F)\|_{L^p_\omega} \approx \|\mathcal{N}F\|_{L^p_\omega} \lesssim \|f + H^L f\|_{L^p_\omega} \lesssim \|f\|_{L^p_\omega} \approx \|\mathcal{N}u\|_{L^p_\omega}.$$

Exercise. Use the results of the last two chapters to give a simple proof of the Fefferman-Stein ([**FS**]) characterization of the Hardy space $H^1(\mathbb{R}^n)$. More specifically, prove that if R_j is the j-th Riesz transform in $\mathbb{R}^n$ and P_t denotes the usual Poisson kernel for $\mathbb{R}^{n+1}_+$, then a complex valued function $f \in L^1(\mathbb{R}^n)$ has $\mathcal{N}(P_t * f) \in L^1(\mathbb{R}^n)$ if and only if $R_j f \in L^1(\mathbb{R}^n)$ for $j = 1, 2, ..., n$.

Hint: Consider $F := \mathcal{C}^L f$ in $\mathbb{R}^{n+1}_+$ and note that F is the harmonic extension of its boundary trace $F|_{\partial \mathbb{R}^{n+1}_+} = \frac{1}{2}(f - \sum_j e_j R_j f)$, i.e. $F = \frac{1}{2} P_t * (f - \sum_j e_j R_j f)$. Now, the direct implication is given by Theorem 5.14, whereas the converse one is seen from Theorem 4.6.

Exercise. Use Lemma 5.16, Theorem 5.14 and Theorem 4.1 to obtain an integral representation formula for arbitrary harmonic functions in $H^p_\omega(\Omega)$.

References

[AJM] Andersson, L., Jawerth, B. and Mitrea, M., *The Cauchy singular integral operator and Clifford wavelets*, in Wavelets: Mathematics and Applications, J. Benedetto and M. Frazier eds. (1993), 519–540.

[AT] Auscher, P. and Tchamitchian, Ph., *Bases d'ondelettes sur des courbes corde-arc, noyau de Cauchy et espaces de Hardy associés*, Rev. Matemática Iberoamericana 5 (1989), 139–170.

[BM] Banuelos, R. and Moore, C., N., *Sharp estimates for the nontangential maximal function and the Lusin area function in Lipschitz domains*, Tran. Amer. Math. Soc. 312 (1989), 641–662.

[Be] Bell, S., R., *The Cauchy Transform, Potential Theory, and Conformal Mapping*, Studies in Advanced Math. Series, C.R.C. Press (1992).

[BDS], Brackx, F., Delanghe R. and Sommen, F., *Clifford Analysis*, Research Notes in Mathematics 76, Pitman Advanced Publishing Company, Boston, London, Melbourne (1982).

[BG] Burkholder, D., L. and Gundy, R., F., *Distribution function inequalities for the area integral*, Studia Math. 44 (1972), 527–544.

[BGS] Burkholder, D., L., Gundy, R., F. and Silverstein, M., L., *A maximal function characterization of the class H^p*, Trans. Amer. Math. Soc. 157 (1971), 137–153.

[Ca] Calderón, A., P., *Boundary value problems in Lipschitz domains*, in Recent progress in Fourier Analysis, Elsevier Science Publishers (1985), 33–48.

[Ch] Christ, M., *Lectures on Singular Integral Operators*, CBMS Reg. Conf. Series in Math. 77 (1990).

[Cl1] Clifford, W., K., *On the classification of Geometric Algebras* (1876).

[Cl2] Clifford, W., K., *Applications of Grassman's extensive algebra*, Amer. J. of Math. 1 (1878), 350–358.

[CF] Coifman, R., R. and Fefferman, C., *Weighted norm inequalities for maximal functions and singular integrals*, Studia Math. 51 (1974), 241–250.

[CJS] Coifman. R., R., Jones, P. and Semmes, S., *Two elementary proofs of the L^2 boundedness of the Cauchy integrals on Lipschitz curves*, Journal of A.M.S. **2** (1989), 553–564.

[CMM] Coifman, R., R., McIntosh, A. and Meyer, Y., *L'intégrale de Cauchy définie un opérateur borné sur L^2 pour les courbes lipschitziennes*, Annals of Math. **116** (1982), 361–387.

[CW] Coifman, R., R. and Weiss, G., *Analyse harmonique non-commutative sur certains espaces homogènes*, Lecture Notes in Mathematics, Springer-Verlag **242** (1971).

[Dah1] Dahlberg, B., J., E., *On estimates of harmonic measure*, Arch. Rat. Mech. Anal. **65** (1977), 275–288.

[Dah2] Dahlberg, B., J., E., *Weighted norm inequalities for the Lusin area integral and the nontangential maximal function for functions harmonic in a Lipschitz domain*, Studia Math. **67** (1980), 297–314.

[Dah3] Dahlberg, B., J., E., *Harmonic functions in Lipschitz domains*, Proc. Symp. Pure and Appl. Math., vol. 35, G. Weiss and S. Waigner eds. (1979), 312–322.

[Dav1] David, G., *Opérateurs intégraux singuliers sur certaines courbes du plan complexe*, Ann. Sci. E.N.S. **17** (1984), 157–189.

[Dav2] David, G., *Wavelets and Singular Integrals on Curves and Surfaces*, Lectures Notes, Springer-Verlag **1465** (1991).

[DJ] David, G. and Journé, J.-L., *A boundedness criterion for generalized Calderón-Zygmund operators*, Annals of Mathematics **120** (1984), 371–379.

[DJS] David, G., Journé, J.-L. and Semmes, S., *Operateurs de Calderón-Zygmund, fonctiones para-accretives et interpolation*, Rev. Mat. Iberoamericana 1 (1985), 1–56.

[Di] Dixon, A., C., *On the Newtonian potential*, Quarterly J. of Math. **35** (1904), 283–296.

[FJR] Fabes, E., Jodeit, M. and Riviére, N., *Potential techniques for boundary value problems on C^1 domains*, Acta Math. **141** (1978), 165–186.

[Fe] Fefferman, C., *Harmonic analysis and H^p spaces*, Studies in Harmonic Analysis (M. Ash editor), M. A. A. Studies in Math. **13** (1976), 38–75.

[FS] Fefferman, C. and Stein, E., M., *H^p Spaces of several variables*, Acta Math. **129**

(1972), 137–193.

[Fu] Fueter, R., *Analytische Funktionen einer Quaternionenvariablen*, Comment. Math. Helv. **4** (1932), 9–20.

[Ga1] Garcia-Cuerva, J., *Weighted Hardy spaces*, Proc. Symp. Pure Math., Providence R. I. **XXXV(1)** (1979), 253–261.

[Ga2] Garcia-Cuerva, J., *Weighted H^p spaces*, Dissertationes Mathematicae **162** (1979).

[GR] Garcia-Cuerva, J. and Rubio de Francia, J., L., *Weighted norm inequalities and related topics*, Mathematical Studies, North-Holland Amsterdam **116** (1985).

[GLQ] Gaudry, G., I., Long, R., L. and Qian, T., *A Martingale Proof of L^2 Boundedness of Clifford-Valued Singular Integrals*, preprint (1991).

[GM1] Gilbert, J. and Murray, M., A., *H^p-Theory on Euclidean space and the Dirac operator*, Rev. Mat. Iberoamericana **4** (1988), 253–289.

[GM2] Gilbert, J. and Murray, M., A., *Clifford Algebras and Dirac Operators in Harmonic Analysis*, Cambridge Studies in Advanced Mathematics **26** (1991).

[Gr] Grisvard, P., *Elliptic Problems in Nonsmooth Domains*, Pitman Advanced Publishing Program (1985).

[GW] Gundy, R., F. and Wheeden, R., L., *Weighted integral inequalities for the nontangential maximal function, Lusin area integral and Walsh-Paley series*, Studia Math. **49** (1974), 107–124.

[HS] Hestenes, D. and Sobczyk, G., *Clifford Algebra to Geometric Calculus. A Unified Language for Mathematics and Physics*, D. Reidel Publ. Comp., Dordrecht, Boston, Lancaster (1984).

[Hf] Hoffman, K., *Banach Spaces of Analytic Functions*, Dover Publ. Inc., New York (1988).

[If] Iftimie, V., *Fonctiones hypercomplexes*, Bull. Math. de la Soc. Sci. Math. de la R. S. Roumanie **4** (1965), 279–332.

[JK] Jerison, D. and Kenig, C., E., *The Neumann problem on Lipschitz domains*, Bull. A.M.S. **4** (1981), 203–207.

[Jo] Journé, J.-L., *Calderón-Zygmund Operators, Pseudo-Differential Operators and the Cauchy Integral of Calderón*, Lecture Notes in Mathematics, Springer-Verlag, Berlin **994** (1983).

[Ju] Jurchescu, M., personal communication.

[JM] Jurchescu, M. and Mitrea, M., *Pluridimensional absolute continuity for differential forms and Stokes formula*, preprint (1993).

[Ka] Kato, T., *Theory for Linear Operators*, Springer-Verlag, Berlin-Heidelberg-New York (1976).

[Ke] Kenig, C., E., *Weighted H^p spaces on Lipschitz domains*, Amer. J. Math. **102** (1980), 129–163.

[KP] Kenig, C., E. and Pipher, J., *Hardy spaces and the Dirichlet problem on Lipschitz domains*, Rev. Mat. Iberoamericana **3** (1987), 191–247.

[Ko1] Koosis, P., *Sommabilité de la fonction maximale et appartenance à H_1*, C. R. Acad. Sc. Paris **286** (1978), 1041-1044.

[Ko2] Koosis, P., *Sommabilité de la fonction maximale et appartenance à H_1. Cas de plusieurs variables*, C. R. Acad. Sc. Paris **288** (1979), 489–492.

[Kr] Krylov, V., I., *On functions regular in a half-plane*, Math. Sb. **6(48)** (1939), 95–138.

[LMS] Li, C., McIntosh, A. and Semmes, S., *Convolution singular integrals on Lipschitz surfaces*, J. Amer. Math. Soc. **5** (1992), 455–481.

[LMQ] Li, C., McIntosh, A. and Qian, T., *Clifford algebras, Fourier transforms, and singular convolution operators on Lipschitz surfaces*, Macquarie Mathematics Report No. 91–087 (1991).

[Mc] McIntosh, A., *Clifford algebras and the higher dimensional Cauchy integral*, Approximation Theory and Function Spaces, Banach Center Publications **22** (1989), 253–267.

[Mc1] McIntosh, A., personal communication.

[MM] McIntosh, A. and Meyer, Y., *Algebres d'opérateurs definis par des integrales singulieres*, C.R. Acad. Sci. Paris, Ser. I Math. **301** (1985), 395–397.

[Me] Meyer, Y., *Ondelettes et opérateurs, I-III*, Hermann, Paris (1990).

[Mi1] Mitrea, M., Ph. D. Thesis.

[Mi2] Mitrea, M., *Hardy spaces and Clifford algebras*, preprint (1991).

[Mi3] Mitrea, M., *Boundary value problems and Hardy spaces associated to the Helmholtz equation on Lipschitz domains*, submitted (1992).

[Mi4] Mitrea, M., *A hypercomplex variable proof of a Burkholder-Gundy-Silverstein type result on Lipschitz domains*, Bull. of the Belgian Math. Soc., Series B **45** (1993), 125–135.

[Mi5] Mitrea, M., *Clifford algebras and boundary estimates for harmonic functions*, in Clifford Algebras and Their Applications in Mathematical Physics, Brackx, F., Delanghe, R. and Serras, H. eds., Kluwer Academic Publishers, Dordrecht (1993), 151-159.

[Mi6] Mitrea, M., *The regularity of the Cauchy integral and related operators on Lipschitz hypersurfaces*, to appear in Complex Variables: Theory and Applications (1993).

[Mi7] Mitrea, M., *Hypercomplex variable techniques in Harmonic Analysis*, to appear in the Proceedings of the Conference on Clifford Algebras in Analysis, Fayetteville, Arkansas, in the Studies in Advanced Mathematics series of C.R.C. Press.

[Mi8] Mitrea, M., *Harmonic and Clifford analytic functions in nonsmooth domains*, submitted (1993).

[Mi9] Mitrea, M., *On Dahlberg's Lusin area integral theorem*, to appear in Proc. of Amer. Math. Soc. (1993).

[MS] Mitrea M. and Sabac, F., *Pompeiu's integral representation formula. History and Mathematics*, preprint (1993).

[Mo1] Moisil, G., C., *Sur l'équation $\triangle u = 0$*, C. R. Acad. Sci. Paris **191** (1930), 984–986.

[Mo2] Moisil, G., C., *Sur les systèmes d'équations de M. Dirac, du type elliptique*, C. R. Acad. Sci. Paris **191** (1930), 1292–1293.

[Mo3] Moisil, G., C., *Sur la généralisation des fonctions conjuguées*, Rendiconti della Reale Accad. Naz. dei Lincei **14** (1931), 401–408.

[MT] Moisil G., C. and Teodorescu, N., *Fonctions holomorphes dans l'espace*, Mathematicae Cluj **5** (1931), 142–150.

[Mu] Murray, M., A., M., *The Cauchy integral, Calderón commutators and conjugations of singular integrals*, Trans. Amer. Math. Soc. **289** (1985), 497–518.

[Ne] Nečas, J., *Les méthodes directes en théorie des équations élliptiques*, Academia, Prague (1967).

[PW] Payne, L. and Weinberger, H., *New bounds in harmonic and biharmonic problems*, J. Math. Phys. **33** (1954), 291–307.

[Po1] Pompeiu, D., *Sur une classe de fonctions d'une variable complexe*, Rendiconti del Circolo Matematico di Palermo **33** (1912), 108–113.

[Po2] Pompeiu, D., *Sur une classe de fonctions d'une variable complexe et sur certaines équations intégrales*, Rendiconti del Circolo Matematico di Palermo **35** (1913), 277–281.

[Po3] Pompeiu, D., *Sur une définition des fonctions holomorphes*, C. R. Acad. Sci. Paris **166** (1918), 209–212.

[Re] Rellich, F., *Darstellung der eigenwerte von $\triangle u + \lambda u$ durch ein Radintegral*, Math. Zeit. **46** (1940), 635–646.

[S] Segovia, C., *On the area function of Lusin*, Studia Math. **33** (1969), 312–343.

[Se1] Semmes, S., *A criterion for the boundedness of singular integrals on hypersurfaces*, Trans. Amer. Math. Soc. **311** (1989), 501–513.

[Se2] Semmes, S., *Square function estimates and the $T(b)$ theorem*, Proc. of A.M.S. **110** (1990), 721–726.

[Se3] Semmes, S., *Differentiable function theory on hypersurfaces (without bounds on their smoothness) in $\mathbb{R}^n$*, Indiana Univ. Math. J. **39** (1990), 985–1004.

[Se4] Semmes, S., *Analysis versus Geometry on a class of rectifiable hypersurfaces in $\mathbb{R}^n$*, Indiana Univ. Math. J. **39** (1990), 1005–1035.

[St] Stein, E., M., *Singular Integrals and Differentiability Properties of Functions*, Princeton University Press, Princeton, N. J. (1970).

[SW1] Stein, E., M. and Weiss, G., *On the theory of harmonic functions of several variables, I. The theory of H^p spaces.*, Acta Math. **103** (1960), 25–62.

[SW2] Stein, E., M. and Weiss, G., *Generalization of the Cauchy-Riemann equations and representations of the rotation group*, Amer. J. Math. **90** (1968), 163–196.

[SW3] Stein, E., M. and Weiss, G., *Introduction to Fourier analysis on Euclidean spaces*, Princeton University Press, Princeton (1971).

[ST] Strömberg, J.-O. and Torchinsky, A., *Weighted Hardy Spaces*, Lecture-Notes in Math., Springer-Verlag, Berlin / Heidelberg **1381** (1989).

[Ta] Tao, T., *Convolution operators generated by right-monogenic kernels*, preprint (1992).

[Tc] Tchamitchian, Ph., *Ondelettes et intégrale de Cauchy sur les courbes lipschitziennes*, Ann. of Math. **129** (1989), 641–649.

[Te] Teodorescu, N., *La Dérivée Areolaire*, Ann. Roumanine des Math., Cahier 3, Bucharest (1936).

[To] Torchinsky, A., *Real-Variable Methods in Harmonic Analysis*, Academic Press, Inc. (1986).

[Ve] Verchota, G., *Layer potentials and regularity for the Dirichlet problem for Laplace's equation in Lipschitz domains*, Journal of Functional Analysis **59** (1984), 572–611.

[Zy] Zygmund, A., *Trigonometric series*, Warsaw (1935).

Notational Index

e_j, e_I: 1,2

$\mathbb{R}_{(n)}$, $\mathbb{C}_{(n)}$: 1,4

D, $\bar{D}$: 5,8,9

$\approx$, $\lesssim$: 5

E_k^L, E_k^R: 35

C_δ^L, C_δ^R: 43

H_δ^L, H_δ^R: 44,45

Σ: 42

$N(y)$, $n(y)$: 42

$\langle \cdot, \cdot \rangle_b$: 53

$\mathcal{H}_\omega^p(\Omega_\pm)$, $\mathcal{K}_\omega^p(\Omega_\pm)$: 61

F_{rad}: 61

F^+: 62

$\mathcal{A}(F)$, $g(F)$: 73

∇_Σ: 82

$H_\omega^p(\Omega)$, $H_\omega^{p,*}(\Omega)$, $H_\omega^{p,1}(\Omega)$: 87,88

∇_T: 89

Re: 4

$V_{(n)}$, $\| \cdot \|_{(n)}$: 11,12

$E(X)$: 9

$\mathcal{F}_k$: 30

$\triangle_k^L$, $\triangle_k^R$: 35

$\Theta_{Q,i}^L$, $\Theta_{Q,i}^R$: 33

H^L, H^R: 45

dS: 61

$\langle \cdot, \cdot \rangle_\Sigma$: 43,71

Γ_α, $\mathcal{N}F$: 61

$\mathcal{C}^L$, $\mathcal{C}^R$: 61

$\mathcal{B}_\pm^L$, $\mathcal{B}_\pm^R$: 70

f^*: 63

R_j, P_{x_0}: 81

$\mathcal{H}_\omega^{p,*}(\Omega)$, $\mathcal{H}_\omega^{p,1}(\Omega)$: 83

$\mathcal{D}f$, $\mathcal{K}f$: 88

$F_\pm$: 91

113

Subject Index

accretive

  -δ-   17,24,25,27

  -pseudo-   30,31,41,53,55

area-function (Lusin)   73,75,78,81,87,103

BMO 37,54,55

Calderón's decomposition   71

Calderón-Zygmund

  -decomposition   39

  -operators   45

  -method of rotation   42

Carleson sequence   37

Cauchy

  -kernel   9,54,72,74,80

  -reproducing formula   11,65,76

  -vanishing theorem   11,47,76,90

  -integral operator   43,44,61,68,70,82,85

Clifford

  -algebra   1,5,16,42,61

  -bilinear form   17,19

  -differentiation   5

  -functional   12,14,54

  -group   3,33,40

  -module   11,15,53

  -Multiresolution Analysis   18,19,27

  -vectors   3,4,42

  -wavelets   16

Coifman, McIntosh and Meyer's theorem   42

Conditional expectation operators   35

Cotlar's inequality	63
Dahlberg's area theorem	104
Dirichlet problem	87,94,95,98
double layer potential operator	88,89,100
dual pair (of wavelet bases)	21,27,30
dyadic cubes	30
Haar Clifford wavelets	30,43,55
Hardy spaces	57,60-62,68-70,70,78,83-85,98
Hardy-Littlewood maximal operator	45,63,99
Hilbert transform	44,45,63,71,86,89
Homogeneous type (spaces of)	41
jump-relations (Plemelj)	71,77,88
Lipschitz	
-domain	61,87,93,103
-function	56,61
Littlewood-Paley	
-g-function	73,75,78,81,98,99,103
-theory	75
monogenic (left, right, two-sided)	9,61,70,73,78,83,90,91,99
Muckenhoupt class	56,62,69
nontangential	
-boundary trace	45,62,70,84,94
-maximal function	61,62,70,87,94,99,103
Neumann problem	94,96,98
Pompeiu's formula	10
principal value	45,88
radial maximal function	61,70,99,103
real part	4,91,103
regularity	
-of the Cauchy operator	82
-of a CMRA	27
-problem	94,96

Riesz
  -basis (left, right)   20,24,33
  -transforms   81,104
Schur's lemma   44
sesquilinear form   14
single layer potential operator   89
square functions   73,104
standard kernel   53
system of conjugate harmonic functions   10,60
T(1) theorem   29
T(b) theorem   53,55,74,80,85
weak boundedness property   54

Lecture Notes in Mathematics

For information about Vols. 1–1394
please contact your bookseller or Springer-Verlag

Vol. 1395: K. Alladi (Ed.), Number Theory, Madras 1987. Proceedings. VII, 234 pages. 1989.

Vol. 1396: L. Accardi, W. von Waldenfels (Eds.), Quantum Probability and Applications IV. Proceedings, 1987. VI, 355 pages. 1989.

Vol. 1397: P.R. Turner (Ed.), Numerical Analysis and Parallel Processing. Seminar, 1987. VI, 264 pages. 1989.

Vol. 1398: A.C. Kim, B.H. Neumann (Eds.), Groups – Korea 1988. Proceedings. V, 189 pages. 1989.

Vol. 1399: W.-P. Barth, H. Lange (Eds.), Arithmetic of Complex Manifolds. Proceedings, 1988. V, 171 pages. 1989.

Vol. 1400: U. Jannsen. Mixed Motives and Algebraic K-Theory. XIII, 246 pages. 1990.

Vol. 1401: J. Steprans, S. Watson (Eds.), Set Theory and its Applications. Proceedings, 1987. V, 227 pages. 1989.

Vol. 1402: C. Carasso, P. Charrier, B. Hanouzet, J.-L. Joly (Eds.), Nonlinear Hyperbolic Problems. Proceedings, 1988. V, 249 pages. 1989.

Vol. 1403: B. Simeone (Ed.), Combinatorial Optimization. Seminar, 1986. V, 314 pages. 1989.

Vol. 1404: M.-P. Malliavin (Ed.), Séminaire d´Algèbre Paul Dubreil et Marie-Paul Malliavin. Proceedings, 1987–1988. IV, 410 pages. 1989.

Vol. 1405: S. Dolecki (Ed.), Optimization. Proceedings, 1988. V, 223 pages. 1989. Vol. 1406: L. Jacobsen (Ed.), Analytic Theory of Continued Fractions III. Proceedings, 1988. VI, 142 pages. 1989.

Vol. 1407: W. Pohlers, Proof Theory. VI, 213 pages. 1989.

Vol. 1408: W. Lück, Transformation Groups and Algebraic K-Theory. XII, 443 pages. 1989.

Vol. 1409: E. Hairer, Ch. Lubich, M. Roche. The Numerical Solution of Differential-Algebraic Systems by Runge-Kutta Methods. VII, 139 pages. 1989.

Vol. 1410: F.J. Carreras, O. Gil-Medrano, A.M. Naveira (Eds.), Differential Geometry. Proceedings, 1988. V, 308 pages. 1989.

Vol. 1411: B. Jiang (Ed.), Topological Fixed Point Theory and Applications. Proceedings. 1988. VI, 203 pages. 1989.

Vol. 1412: V.V. Kalashnikov, V.M. Zolotarev (Eds.), Stability Problems for Stochastic Models. Proceedings, 1987. X, 380 pages. 1989.

Vol. 1413: S. Wright, Uniqueness of the Injective III₁Factor. III, 108 pages. 1989.

Vol. 1414: E. Ramirez de Arellano (Ed.), Algebraic Geometry and Complex Analysis. Proceedings, 1987. VI, 180 pages. 1989.

Vol. 1415: M. Langevin, M. Waldschmidt (Eds.), Cinquante Ans de Polynômes. Fifty Years of Polynomials. Proceedings, 1988. IX, 235 pages.1990.

Vol. 1416: C. Albert (Ed.), Géométrie Symplectique et Mécanique. Proceedings, 1988. V, 289 pages. 1990.

Vol. 1417: A.J. Sommese, A. Biancofiore, E.L. Livorni (Eds.), Algebraic Geometry. Proceedings, 1988. V, 320 pages. 1990.

Vol. 1418: M. Mimura (Ed.), Homotopy Theory and Related Topics. Proceedings, 1988. V, 241 pages. 1990.

Vol. 1419: P.S. Bullen, P.Y. Lee, J.L. Mawhin, P. Muldowney, W.F. Pfeffer (Eds.), New Integrals. Proceedings, 1988. V, 202 pages. 1990.

Vol. 1420: M. Galbiati, A. Tognoli (Eds.), Real Analytic Geometry. Proceedings, 1988. IV, 366 pages. 1990.

Vol. 1421: H.A. Biagioni, A Nonlinear Theory of Generalized Functions, XII, 214 pages. 1990.

Vol. 1422: V. Villani (Ed.), Complex Geometry and Analysis. Proceedings, 1988. V, 109 pages. 1990.

Vol. 1423: S.O. Kochman, Stable Homotopy Groups of Spheres: A Computer-Assisted Approach. VIII, 330 pages. 1990.

Vol. 1424: F.E. Burstall, J.H. Rawnsley, Twistor Theory for Riemannian Symmetric Spaces. III, 112 pages. 1990.

Vol. 1425: R.A. Piccinini (Ed.), Groups of Self-Equivalences and Related Topics. Proceedings, 1988. V, 214 pages. 1990.

Vol. 1426: J. Azéma, P.A. Meyer, M. Yor (Eds.), Séminaire de Probabilités XXIV, 1988/89. V, 490 pages. 1990.

Vol. 1427: A. Ancona, D. Geman, N. Ikeda, École d'Eté de Probabilités de Saint Flour XVIII, 1988. Ed.: P.L. Hennequin. VII, 330 pages. 1990.

Vol. 1428: K. Erdmann, Blocks of Tame Representation Type and Related Algebras. XV. 312 pages. 1990.

Vol. 1429: S. Homer, A. Nerode, R.A. Platek, G.E. Sacks, A. Scedrov, Logic and Computer Science. Seminar, 1988. Editor: P. Odifreddi. V, 162 pages. 1990.

Vol. 1430: W. Bruns, A. Simis (Eds.), Commutative Algebra. Proceedings. 1988. V, 160 pages. 1990.

Vol. 1431: J.G. Heywood, K. Masuda, R. Rautmann, V.A. Solonnikov (Eds.), The Navier-Stokes Equations – Theory and Numerical Methods. Proceedings, 1988. VII, 238 pages. 1990.

Vol. 1432: K. Ambos-Spies, G.H. Müller, G.E. Sacks (Eds.), Recursion Theory Week. Proceedings, 1989. VI, 393 pages. 1990.

Vol. 1433: S. Lang, W. Cherry, Topics in Nevanlinna Theory. II, 174 pages.1990.

Vol. 1434: K. Nagasaka, E. Fouvry (Eds.), Analytic Number Theory. Proceedings, 1988. VI, 218 pages. 1990.

Vol. 1435: St. Ruscheweyh, E.B. Saff, L.C. Salinas, R.S. Varga (Eds.), Computational Methods and Function Theory. Proceedings, 1989. VI, 211 pages. 1990.

Vol. 1436: S. Xambó-Descamps (Ed.), Enumerative Geometry. Proceedings, 1987. V, 303 pages. 1990.

Vol. 1437: H. Inassaridze (Ed.), K-theory and Homological Algebra. Seminar, 1987–88. V, 313 pages. 1990.

Vol. 1438: P.G. Lemarié (Ed.) Les Ondelettes en 1989. Seminar. IV, 212 pages. 1990.

Vol. 1439: E. Bujalance, J.J. Etayo, J.M. Gamboa, G. Gromadzki. Automorphism Groups of Compact Bordered Klein Surfaces: A Combinatorial Approach. XIII, 201 pages. 1990.

Vol. 1440: P. Latiolais (Ed.), Topology and Combinatorial Groups Theory. Seminar, 1985–1988. VI, 207 pages. 1990.

Vol. 1441: M. Coornaert, T. Delzant, A. Papadopoulos. Géométrie et théorie des groupes. X, 165 pages. 1990.

Vol. 1442: L. Accardi, M. von Waldenfels (Eds.), Quantum Probability and Applications V. Proceedings, 1988. VI, 413 pages. 1990.

Vol. 1443: K.H. Dovermann, R. Schultz, Equivariant Surgery Theories and Their Periodicity Properties. VI, 227 pages. 1990.

Vol. 1444: H. Korezlioglu, A.S. Ustunel (Eds.), Stochastic Analysis and Related Topics VI. Proceedings, 1988. V, 268 pages. 1990.

Vol. 1445: F. Schulz, Regularity Theory for Quasilinear Elliptic Systems and – Monge Ampère Equations in Two Dimensions. XV, 123 pages. 1990.

Vol. 1446: Methods of Nonconvex Analysis. Seminar, 1989. Editor: A. Cellina. V, 206 pages. 1990.

Vol. 1447: J.-G. Labesse, J. Schwermer (Eds), Cohomology of Arithmetic Groups and Automorphic Forms. Proceedings, 1989. V, 358 pages. 1990.

Vol. 1448: S.K. Jain, S.R. López-Permouth (Eds.), Non-Commutative Ring Theory. Proceedings, 1989. V, 166 pages. 1990.

Vol. 1449: W. Odyniec, G. Lewicki, Minimal Projections in Banach Spaces. VIII, 168 pages. 1990.

Vol. 1450: H. Fujita, T. Ikebe, S.T. Kuroda (Eds.), Functional-Analytic Methods for Partial Differential Equations. Proceedings, 1989. VII, 252 pages. 1990.

Vol. 1451: L. Alvarez-Gaumé, E. Arbarello, C. De Concini, N.J. Hitchin, Global Geometry and Mathematical Physics. Montecatini Terme 1988. Seminar. Editors: M. Francaviglia, F. Gherardelli. IX, 197 pages. 1990.

Vol. 1452: E. Hlawka, R.F. Tichy (Eds.), Number-Theoretic Analysis. Seminar, 1988–89. V, 220 pages. 1990.

Vol. 1453: Yu.G. Borisovich, Yu.E. Gliklikh (Eds.), Global Analysis – Studies and Applications IV. V, 320 pages. 1990.

Vol. 1454: F. Baldassari, S. Bosch, B. Dwork (Eds.), p-adic Analysis. Proceedings, 1989. V, 382 pages. 1990.

Vol. 1455: J.-P. Françoise, R. Roussarie (Eds.), Bifurcations of Planar Vector Fields. Proceedings, 1989. VI, 396 pages. 1990.

Vol. 1456: L.G. Kovács (Ed.), Groups – Canberra 1989. Proceedings. XII, 198 pages. 1990.

Vol. 1457: O. Axelsson, L.Yu. Kolotilina (Eds.), Preconditioned Conjugate Gradient Methods. Proceedings, 1989. V, 196 pages. 1990.

Vol. 1458: R. Schaaf, Global Solution Branches of Two Point Boundary Value Problems. XIX, 141 pages. 1990.

Vol. 1459: D. Tiba, Optimal Control of Nonsmooth Distributed Parameter Systems. VII, 159 pages. 1990.

Vol. 1460: G. Toscani, V. Boffi, S. Rionero (Eds.), Mathematical Aspects of Fluid Plasma Dynamics. Proceedings, 1988. V, 221 pages. 1991.

Vol. 1461: R. Gorenflo, S. Vessella, Abel Integral Equations. VII, 215 pages. 1991.

Vol. 1462: D. Mond, J. Montaldi (Eds.), Singularity Theory and its Applications. Warwick 1989, Part I. VIII, 405 pages. 1991.

Vol. 1463: R. Roberts, I. Stewart (Eds.), Singularity Theory and its Applications. Warwick 1989, Part II. VIII, 322 pages. 1991.

Vol. 1464: D. L. Burkholder, E. Pardoux, A. Sznitman, Ecole d'Eté de Probabilités de Saint- Flour XIX-1989. Editor: P. L. Hennequin. VI, 256 pages. 1991.

Vol. 1465: G. David, Wavelets and Singular Integrals on Curves and Surfaces. X, 107 pages. 1991.

Vol. 1466: W. Banaszczyk, Additive Subgroups of Topological Vector Spaces. VII, 178 pages. 1991.

Vol. 1467: W. M. Schmidt, Diophantine Approximations and Diophantine Equations. VIII, 217 pages. 1991.

Vol. 1468: J. Noguchi, T. Ohsawa (Eds.), Prospects in Complex Geometry. Proceedings, 1989. VII, 421 pages. 1991.

Vol. 1469: J. Lindenstrauss, V. D. Milman (Eds.), Geometric Aspects of Functional Analysis. Seminar 1989-90. XI, 191 pages. 1991.

Vol. 1470: E. Odell, H. Rosenthal (Eds.), Functional Analysis. Proceedings, 1987-89. VII, 199 pages. 1991.

Vol. 1471: A. A. Panchishkin, Non-Archimedean L-Functions of Siegel and Hilbert Modular Forms. VII, 157 pages. 1991.

Vol. 1472: T. T. Nielsen, Bose Algebras: The Complex and Real Wave Representations. V, 132 pages. 1991.

Vol. 1473: Y. Hino, S. Murakami, T. Naito, Functional Differential Equations with Infinite Delay. X, 317 pages. 1991.

Vol. 1474: S. Jackowski, B. Oliver, K. Pawałowski (Eds.), Algebraic Topology, Poznań 1989. Proceedings. VIII, 397 pages. 1991.

Vol. 1475: S. Busenberg, M. Martelli (Eds.), Delay Differential Equations and Dynamical Systems. Proceedings, 1990. VIII, 249 pages. 1991.

Vol. 1476: M. Bekkali, Topics in Set Theory. VII, 120 pages. 1991.

Vol. 1477: R. Jajte, Strong Limit Theorems in Noncommutative L_2-Spaces. X, 113 pages. 1991.

Vol. 1478: M.-P. Malliavin (Ed.), Topics in Invariant Theory. Seminar 1989-1990. VI, 272 pages. 1991.

Vol. 1479: S. Bloch, I. Dolgachev, W. Fulton (Eds.), Algebraic Geometry. Proceedings, 1989. VII, 300 pages. 1991.

Vol. 1480: F. Dumortier, R. Roussarie, J. Sotomayor, H. Żołądek, Bifurcations of Planar Vector Fields: Nilpotent Singularities and Abelian Integrals. VIII, 226 pages. 1991.

Vol. 1481: D. Ferus, U. Pinkall, U. Simon, B. Wegner (Eds.), Global Differential Geometry and Global Analysis. Proceedings, 1991. VIII, 283 pages. 1991.

Vol. 1482: J. Chabrowski, The Dirichlet Problem with L^2-Boundary Data for Elliptic Linear Equations. VI, 173 pages. 1991.

Vol. 1483: E. Reithmeier, Periodic Solutions of Nonlinear Dynamical Systems. VI, 171 pages. 1991.

Vol. 1484: H. Delfs, Homology of Locally Semialgebraic Spaces. IX, 136 pages. 1991.

Vol. 1485: J. Azéma, P. A. Meyer, M. Yor (Eds.), Séminaire de Probabilités XXV. VIII, 440 pages. 1991.

Vol. 1486: L. Arnold, H. Crauel, J.-P. Eckmann (Eds.), Lyapunov Exponents. Proceedings, 1990. VIII, 365 pages. 1991.

Vol. 1487: E. Freitag, Singular Modular Forms and Theta Relations. VI, 172 pages. 1991.

Vol. 1488: A. Carboni, M. C. Pedicchio, G. Rosolini (Eds.), Category Theory. Proceedings, 1990. VII, 494 pages. 1991.

Vol. 1489: A. Mielke, Hamiltonian and Lagrangian Flows on Center Manifolds. X, 140 pages. 1991.

Vol. 1490: K. Metsch, Linear Spaces with Few Lines. XIII, 196 pages. 1991.

Vol. 1491: E. Lluis-Puebla, J.-L. Loday, H. Gillet, C. Soulé, V. Snaith, Higher Algebraic K-Theory: an overview. IX, 164 pages. 1992.

Vol. 1492: K. R. Wicks, Fractals and Hyperspaces. VIII, 168 pages. 1991.

Vol. 1493: E. Benoît (Ed.), Dynamic Bifurcations. Proceedings, Luminy 1990. VII, 219 pages. 1991.

Vol. 1494: M.-T. Cheng, X.-W. Zhou, D.-G. Deng (Eds.), Harmonic Analysis. Proceedings, 1988. IX, 226 pages. 1991.

Vol. 1495: J. M. Bony, G. Grubb, L. Hörmander, H. Komatsu, J. Sjöstrand, Microlocal Analysis and Applications. Montecatini Terme, 1989. Editors: L. Cattabriga, L. Rodino. VII, 349 pages. 1991.

Vol. 1496: C. Foias, B. Francis, J. W. Helton, H. Kwakernaak, J. B. Pearson, H_∞-Control Theory. Como, 1990. Editors: E. Mosca, L. Pandolfi. VII, 336 pages. 1991.

Vol. 1497: G. T. Herman, A. K. Louis, F. Natterer (Eds.), Mathematical Methods in Tomography. Proceedings 1990. X, 268 pages. 1991.

Vol. 1498: R. Lang, Spectral Theory of Random Schrödinger Operators. X, 125 pages. 1991.

Vol. 1499: K. Taira, Boundary Value Problems and Markov Processes. IX, 132 pages. 1991.

Vol. 1500: J.-P. Serre, Lie Algebras and Lie Groups. VII, 168 pages. 1992.

Vol. 1501: A. De Masi, E. Presutti, Mathematical Methods for Hydrodynamic Limits. IX, 196 pages. 1991.

Vol. 1502: C. Simpson, Asymptotic Behavior of Monodromy. V, 139 pages. 1991.

Vol. 1503: S. Shokranian, The Selberg-Arthur Trace Formula (Lectures by J. Arthur). VII, 97 pages. 1991.

Vol. 1504: J. Cheeger, M. Gromov, C. Okonek, P. Pansu, Geometric Topology: Recent Developments. Editors: P. de Bartolomeis, F. Tricerri. VII, 197 pages. 1991.

Vol. 1505: K. Kajitani, T. Nishitani, The Hyperbolic Cauchy Problem. VII, 168 pages. 1991.

Vol. 1506: A. Buium, Differential Algebraic Groups of Finite Dimension. XV, 145 pages. 1992.

Vol. 1507: K. Hulek, T. Peternell, M. Schneider, F.-O. Schreyer (Eds.), Complex Algebraic Varieties. Proceedings, 1990. VII, 179 pages. 1992.

Vol. 1508: M. Vuorinen (Ed.), Quasiconformal Space Mappings. A Collection of Surveys 1960-1990. IX, 148 pages. 1992.

Vol. 1509: J. Aguadé, M. Castellet, F. R. Cohen (Eds.), Algebraic Topology - Homotopy and Group Cohomology. Proceedings, 1990. X, 330 pages. 1992.

Vol. 1510: P. P. Kulish (Ed.), Quantum Groups. Proceedings, 1990. XII, 398 pages. 1992.

Vol. 1511: B. S. Yadav, D. Singh (Eds.), Functional Analysis and Operator Theory. Proceedings, 1990. VIII, 223 pages. 1992.

Vol. 1512: L. M. Adleman, M.-D. A. Huang, Primality Testing and Abelian Varieties Over Finite Fields. VII, 142 pages. 1992.

Vol. 1513: L. S. Block, W. A. Coppel, Dynamics in One Dimension. VIII, 249 pages. 1992.

Vol. 1514: U. Krengel, K. Richter, V. Warstat (Eds.), Ergodic Theory and Related Topics III, Proceedings, 1990. VIII, 236 pages. 1992.

Vol. 1515: E. Ballico, F. Catanese, C. Ciliberto (Eds.), Classification of Irregular Varieties. Proceedings, 1990. VII, 149 pages. 1992.

Vol. 1516: R. A. Lorentz, Multivariate Birkhoff Interpolation. IX, 192 pages. 1992.

Vol. 1517: K. Keimel, W. Roth, Ordered Cones and Approximation. VI, 134 pages. 1992.

Vol. 1518: H. Stichtenoth, M. A. Tsfasman (Eds.), Coding Theory and Algebraic Geometry. Proceedings, 1991. VIII, 223 pages. 1992.

Vol. 1519: M. W. Short, The Primitive Soluble Permutation Groups of Degree less than 256. IX, 145 pages. 1992.

Vol. 1520: Yu. G. Borisovich, Yu. E. Gliklikh (Eds.), Global Analysis – Studies and Applications V. VII, 284 pages. 1992.

Vol. 1521: S. Busenberg, B. Forte, H. K. Kuiken, Mathematical Modelling of Industrial Process. Bari, 1990. Editors: V. Capasso, A. Fasano. VII, 162 pages. 1992.

Vol. 1522: J.-M. Delort, F. B. I. Transformation. VII, 101 pages. 1992.

Vol. 1523: W. Xue, Rings with Morita Duality. X, 168 pages. 1992.

Vol. 1524: M. Coste, L. Mahé, M.-F. Roy (Eds.), Real Algebraic Geometry. Proceedings, 1991. VIII, 418 pages. 1992.

Vol. 1525: C. Casacuberta, M. Castellet (Eds.), Mathematical Research Today and Tomorrow. VII, 112 pages. 1992.

Vol. 1526: J. Azéma, P. A. Meyer, M. Yor (Eds.), Séminaire de Probabilités XXVI. X, 633 pages. 1992.

Vol. 1527: M. I. Freidlin, J.-F. Le Gall, Ecole d'Eté de Probabilités de Saint-Flour XX – 1990. Editor: P. L. Hennequin. VIII, 244 pages. 1992.

Vol. 1528: G. Isac, Complementarity Problems. VI, 297 pages. 1992.

Vol. 1529: J. van Neerven, The Adjoint of a Semigroup of Linear Operators. X, 195 pages. 1992.

Vol. 1530: J. G. Heywood, K. Masuda, R. Rautmann, S. A. Solonnikov (Eds.), The Navier-Stokes Equations II – Theory and Numerical Methods. IX, 322 pages. 1992.

Vol. 1531: M. Stoer, Design of Survivable Networks. IV, 206 pages. 1992.

Vol. 1532: J. F. Colombeau, Multiplication of Distributions. X, 184 pages. 1992.

Vol. 1533: P. Jipsen, H. Rose, Varieties of Lattices. X, 162 pages. 1992.

Vol. 1534: C. Greither, Cyclic Galois Extensions of Commutative Rings. X, 145 pages. 1992.

Vol. 1535: A. B. Evans, Orthomorphism Graphs of Groups. VIII, 114 pages. 1992.

Vol. 1536: M. K. Kwong, A. Zettl, Norm Inequalities for Derivatives and Differences. VII, 150 pages. 1992.

Vol. 1537: P. Fitzpatrick, M. Martelli, J. Mawhin, R. Nussbaum, Topological Methods for Ordinary Differential Equations. Montecatini Terme, 1991. Editors: M. Furi, P. Zecca. VII, 218 pages. 1993.

Vol. 1538: P.-A. Meyer, Quantum Probability for Probabilists. X, 287 pages. 1993.

Vol. 1539: M. Coornaert, A. Papadopoulos, Symbolic Dynamics and Hyperbolic Groups. VIII, 138 pages. 1993.

Vol. 1540: H. Komatsu (Ed.), Functional Analysis and Related Topics, 1991. Proceedings. XXI, 413 pages. 1993.

Vol. 1541: D. A. Dawson, B. Maisonneuve, J. Spencer, Ecole d´ Eté de Probabilités de Saint-Flour XXI - 1991. Editor: P. L. Hennequin. VIII, 356 pages. 1993.

Vol. 1542: J.Fröhlich, Th.Kerler, Quantum Groups, Quantum Categories and Quantum Field Theory. VII, 431 pages. 1993.

Vol. 1543: A. L. Dontchev, T. Zolezzi, Well-Posed Optimization Problems. XII, 421 pages. 1993.

Vol. 1544: M.Schürmann, White Noise on Bialgebras. VII, 146 pages. 1993.

Vol. 1545: J. Morgan, K. O'Grady, Differential Topology of Complex Surfaces. VIII, 224 pages. 1993.

Vol. 1546: V. V. Kalashnikov, V. M. Zolotarev (Eds.), Stability Problems for Stochastic Models. Proceedings, 1991. VIII, 229 pages. 1993.

Vol. 1547: P. Harmand, D. Werner, W. Werner, M-ideals in Banach Spaces and Banach Algebras. VIII, 387 pages. 1993.

Vol. 1548: T. Urabe, Dynkin Graphs and Quadrilateral Singularities. VI, 233 pages. 1993.

Vol. 1549: G. Vainikko, Multidimensional Weakly Singular Integral Equations. XI, 159 pages. 1993.

Vol. 1550: A. A. Gonchar, E. B. Saff (Eds.), Methods of Approximation Theory in Complex Analysis and Mathematical Physics IV, 222 pages. 1993.

Vol. 1551: L. Arkeryd, P. L. Lions, P.A. Markowich, S.R. S. Varadhan. Nonequilibrium Problems in Many-Particle Systems. Montecatini, 1992. Editors: C. Cercignani, M. Pulvirenti. VII, 158 pages 1993.

Vol. 1552: J. Hilgert, K.-H. Neeb, Lie Semigroups and their Applications. XII, 315 pages. 1993.

Vol. 1553: J.-L- Colliot-Thélène, J. Kato, P. Vojta. Arithmetic Algebraic Geometry. Trento, 1991. Editor: E. Ballico. VII, 223 pages. 1993.

Vol. 1554: A. K. Lenstra, H. W. Lenstra, Jr. (Eds.), The Development of the Number Field Sieve. VIII, 131 pages. 1993.

Vol. 1555: O. Liess, Conical Refraction and Higher Microlocalization. X, 389 pages. 1993.

Vol. 1556: S. B. Kuksin, Nearly Integrable Infinite-Dimensional Hamiltonian Systems. XXVII, 101 pages. 1993.

Vol. 1557: J. Azéma, P. A. Meyer, M. Yor (Eds.), Séminaire de Probabilités XXVII. VI, 327 pages. 1993.

Vol. 1558: T. J. Bridges, J. E. Furter, Singularity Theory and Equivariant Symplectic Maps. VI, 226 pages. 1993.

Vol. 1559: V. G. Sprindžuk, Classical Diophantine Equations. XII, 228 pages. 1993.

Vol. 1560: T. Bartsch, Topological Methods for Variational Problems with Symmetries. X, 152 pages. 1993.

Vol. 1561: I. S. Molchanov, Limit Theorems for Unions of Random Closed Sets. X, 157 pages. 1993.

Vol. 1562: G. Harder, Eisensteinkohomologie und die Konstruktion gemischter Motive. XX, 184 pages. 1993.

Vol. 1563: E. Fabes, M. Fukushima, L. Gross, C. Kenig, M. Röckner, D. W. Stroock, Dirichlet Forms. Varenna, 1992. Editors: G. Dell'Antonio, U. Mosco. VII, 245 pages. 1993.

Vol. 1564: J. Jorgenson, S. Lang, Basic Analysis of Regularized Series and Products. IX, 122 pages. 1993.

Vol. 1565: L. Boutet de Monvel, C. De Concini, C. Procesi, P. Schapira, M. Vergne. D-modules, Representation Theory, and Quantum Groups. Venezia, 1992. Editors: G. Zampieri, A. D'Agnolo. VII, 217 pages. 1993.

Vol. 1566: B. Edixhoven, J.-H. Evertse (Eds.), Diophantine Approximation and Abelian Varieties. XIII, 127 pages. 1993.

Vol. 1567: R. L. Dobrushin, S. Kusuoka, Statistical Mechanics and Fractals. VII, 98 pages. 1993.

Vol. 1568: F. Weisz, Martingale Hardy Spaces and their Application in Fourier Analysis. VIII, 217 pages. 1994.

Vol. 1569: V. Totik, Weighted Approximation with Varying Weight. VI, 117 pages. 1994.

Vol. 1570: R. deLaubenfels, Existence Families, Functional Calculi and Evolution Equations. XV, 234 pages. 1994.

Vol. 1571: S. Yu. Pilyugin, The Space of Dynamical Systems with the C^0-Topology. X, 188 pages. 1994.

Vol. 1572: L. Göttsche, Hilbert Schemes of Zero-Dimensional Subschemes of Smooth Varieties. IX, 196 pages. 1994.

Vol. 1573: V. P. Havin, N. K. Nikolski (Eds.), Linear and Complex Analysis – Problem Book 3 – Part I. XXII, 489 pages. 1994.

Vol. 1574: V. P. Havin, N. K. Nikolski (Eds.), Linear and Complex Analysis – Problem Book 3 – Part II. XXII, 507 pages. 1994.

Vol. 1575: M. Mitrea, Clifford Wavelets, Singular Integrals, and Hardy Spaces. XI, 116 pages. 1994.